*f*P

Also by Victor Davis Hanson

Fields Without Dreams

Warfare and Agriculture in Classical Greece

The Western Way of War

Hoplites: The Ancient Greek Battle Experience, editor

The Other Greeks

Who Killed Homer? (with John Heath)

The Soul of Battle

The Wars of the Ancient Greeks

THE LAND WAS EVERYTHING

EVERYTHING

Letters from
an American Farmer

VICTOR DAVIS HANSON

THE FREE PRESS

NEW YORK LONDON SYDNEY SINGAPORE

THE FREE PRESS
A Division of Simon & Schuster, Inc.
1230 Avenue of the Americas
New York, NY 10020

Designed by Brian Mulligan

Manufactured in the United States of America

10   9   8   7   6   5   4   3   2   1

Library of Congress Cataloging-in-Publication Data

Hanson, Victor Davis.
    The land was everything : letters from an American farmer / Victor Davis Hanson.
        p. cm.
        1. Agriculture—United States.   2. Agriculture—Environmental aspects—
United States.   3. Nature—Effect of human beings on—United States.   4. St. John
de Crèvecoeur, J. Hector, 1735–1813. Letters from an American farmer.   I. Title.

SS441.H28   2000
630'.973—dc21   99-055317

ISBN   0-684-84501-6

In memory of Pauline Davis Hanson

CONTENTS

Part Three
Man versus Self

FOREWORD

There aren't many people in this country who farm anymore. Some people consider this a social and economic victory while others consider it a major indicator of social and economic decline and a big mistake. Certainly it is a shift in our national arrangements that has profound consequences for how we identify and consider ourselves, what we aspire to, and how we organize our lives. It does not seem likely, no matter how fervently anyone feels about this issue, that the majority of our citizenry will return to farming (though many people like to engage with the natural world through gardens, sports, and animal companions, and also prefer exurban life to urban life). It is therefore true that the mental discipline as well as the specific lessons that farming teaches those who engage in it now exist outside the mainstream of American thought and are on the verge of being lost. Victor Davis Hanson, in this volume, *The Land Was Everything,* offers a model of what is being lost. As with any endangered species, the reader of this volume may ask herself or himself, why does this beast seem so strange, and what does it matter to me if it disappears?

We may say, perhaps, that the agricultural revolution of some seven thousand years ago marked the moment when mankind decided to impose itself upon the natural world. To this imposition we now alive undoubtedly owe our existence, because agriculture was what allowed the human species to leap out of its previous

ecological niche and transform every corner of the globe into a potential spot for human habitation. Whatever feelings we have about
this, whether appreciation of the beauties of human culture or regrets at the destruction of nature, these feelings center in the characteristic ambiguity of agriculture and farming. The whole
endeavor is either a realization of the inherent potential of thoughtless brute nature through the application of human reason or a fall
from grace, or both. And on this continent, in the last three hundred years, we have seen the first agricultural revolution recapitu-
lated with greater force and at a higher speed; the native American,
fishing, hunting, and gathering, was replaced by a man with oxen,
then horses, then a tractor, then an armory of machines and chemicals, all in the space of a few generations. It is this quick recapitula-
tion that Hanson considers in *The Land Was Everything*.

The farmer's great trial and his great joy are the same—he engages every day, over and over, with the natural world, and tries to
get it to reproduce itself reliably and abundantly. The engagement
is not innocent nor bucolic nor pastoral, not peaceful nor happy
nor even, at times, productive, but, Hanson maintains, it is basic
and honest, and from it a certain type of citizen grows, who is
skeptical of cant and ease, who values labor and sweat, autonomy
and responsibility. Farmers, Hanson believes, are the original and
essential creators of democracy, and he fears for the future of a
democracy without this originating component.

Victor Davis Hanson is an eloquent and impassioned writer.
He does not promise to charm, but only to pull no punches, and
the picture of farming that we have here is not like any other. In
Hanson's world, the farmer is surrounded by enemies—viruses,
bacteria, fungi, insects, weeds, other farmers, local vagrants of all
types, developers, food conglomerate executives, bankers, and
last but not least, himself. On the other hand, there is the ever-
present mystery of making order of the world, a seductive and

even joyous activity that sometimes results in a terrific crop that sells for a great profit. Hanson's tone is, by turns, angry, bitter, aggressive, stoical, funny, meditative, and rueful. Farming, in California, where Hanson's family grows grapes and fruits, is not about quilting bees and good neighbors helping each other, or staple crops like soybeans, wheat, and corn, but about water and water battles, fashionable and unfashionable specialty crops, immigrant labor, smog and urban sprawl. It is about giving in to the temptation to grow thirsty crops in a place where there is abundant sun and fine soil, but no rainfall most of the year. It is about defying the local ecosystem and remaking the flat plain in the image of human desires, about raising up a certain kind of order that is, inevitably, temporary. The flat, measured, beautiful grids of vines and trees must come and go, because they must either fail to produce and be removed or they must succeed in producing and be replaced by a growing population, drawn to the beauty of the spot that the vines and trees have civilized. Farming in California, like everything in California, is about the cost of pleasure. No one needs raisins, plums, or white peaches to live, nor pinot noir grapes or movies or luxury real estate developments. What people *need* can be grown, built, and seen in the modest Midwest. California is about what we have an appetite for, and what we are willing to pay for it, and, from Hanson's perspective as a farmer, what happens to those whose livelihood depends upon the vagaries of our desires as opposed to the constancy of our needs.

It has often been said, though, that what is happening in California prefigures what will soon happen in the rest of the country, and so even though the contrasts and conflicts of farming in California are sharper than elsewhere, they are not different in kind. As Hanson eloquently shows, farming, like all character-building endeavors, is not fun. Farming, even in the age of tractors and chemicals, requires imagination, persistence, knowledge, precision, and good timing. Even then, when all one's techniques are in place, it can often require the giving up of every dream, when the crop is destroyed at the last moment by weather or disease. Farming, in other words, requires the sort of dogged dedication that is its own reward, because sometimes it has no other reward. To Hanson, this form of dedication is unlike any other, and all the other limbs of society feed off this root, without gratitude, without respect, and even, sometimes, without the simplest knowledge of where sustenance comes from and what it takes to provide it.

Victor Davis Hanson is unique in American letters. He writes not only about farming but about Greek military history, general military history, and classics as a discipline. He is a prolific writer for whom the flow of language seems natural and easily expressive. He teaches classics at the University of California at Fresno. He is young, only in his mid-forties. But, most important, he is an active farmer living in the house he was born in on a farm his family established over a hundred years ago. His voice is educated and eloquent, but also passionate and curmudgeonly. Not for Hanson the safety of cool reason or detached observation. He is, first and foremost, a controversialist, willing to say anything to awaken us from our complacency. Is he a Cassandra, whose prophecies, though true, go unheeded as we around him move steadily toward doom of some sort? Perhaps. But I like to think of him as a goad, a man who is not afraid to start a fight, the bigger the better, because conflict accelerates change and brings people to a more specific consciousness of their goals. Chances are that few will agree wholly with Victor Davis Hanson, chances are that if his opinions don't offend, then his tone might, but that, in the end, is the point, when issues of identity and fate are at stake.

Jane Smiley

RURAL LIFE, ONCE AND FUTURE

The American is a new man, who acts upon new principles; he must therefore entertain new ideas and form new opinions. From involuntary idleness, servile dependence, penury, and useless labour, he has passed to toils of a very different nature rewarded by ample subsistence. This is an American.

—J. Hector St. John de Crèvecoeur,
Letters from an American Farmer

I.

Farmers see things as others do not. Their age-old knowledge is more than the practical experience that comes from the art of growing food or from the independence of rural living. It involves a radically different—often tragic—view of human nature itself that slowly grows through the difficult struggle to work and survive from the land. Destroyed by hail that most others ignore, praying for a rain that few will notice, increasingly foreclosed upon in a sea of cash, smug in their ability to nourish thousands but bewildered that they cannot feed their family, apart from town but dependent on those who are not, still confused over how and why plants usually produce harvests but sometimes do not, the last generation of American farmers have become foreign to their countrymen, who were once as they.

In these letters I wish to pass on some of that social and civic wisdom that I have learned from those now mostly dead or bankrupt, not because their insight necessarily has any archival or antiquarian importance, but because the farmer's understanding of man and society in our present age is absolutely critical to the survival of democracy as we once knew it. Democracy at its inceptions, ancient and American, has always been the outgrowth of an agrarian society; but its old bones now have new and different flesh. Consensual government can continue in the vastly transformed conditions of great wealth, urbanism, and rapidly changing technology never foreseen by its originators; but whether democracy can still instill virtue among its citizens will be answered by the age that is upon us, which for the first time in the history of civilization will at last see a democracy without farmers.

I claim no originality in the idea of writing letters about the farmer's lessons to those who do not farm. Over two hundred years ago J. Hector St. John de Crèvecoeur published a collection of twelve essays on American culture and rural life under the title *Letters from an American Farmer* (1782). Crèvecoeur's letters are generally regarded as the beginning of American literature, inasmuch as they are the first formal expressions of what it was to be "American." The opening to homesteaders of new frontier lands across the eastern seaboard, the immigration and assimilation of a wide variety of Europeans, and the turmoil of the American Revolution convinced Crèvecoeur that he was witnessing at the end of the eighteenth century the birth of a unique nation and a singular man. In his view, freeholding yeomanry lay at the heart of this great experiment in creating a middling, rambunctious democratic citizenry that could not be fooled, enticed, or enslaved—a society quite different even from the emerging consensual govern-

ments in contemporary Europe, which had neither the land nor
the freedom to work it that we enjoyed. In America, the European

> now feels himself a man because he is treated as such; the
> laws of his own country had overlooked him in his insignifi-
> cancy; the laws of this cover him with their mantle. Judge
> what an alteration there must arise in the mind and the
> thoughts of this man. He begins to forget his former servi-
> tude and dependence; his heart involuntarily swells and
> grows; this first swell inspires him with those new thoughts
> which constitute an American. What love can he entertain
> for a country where his existence was a burden to him; if he
> is a generous, good man, the love of this new adoptive parent
> will sink deep into his heart.

Part formal essays, part autobiographical memoirs, part fictive
sketches—on everything from the island of Nantucket to slavery
to the American hummingbird—the letters of Crèvecoeur are
rambling, confused, and at times almost unreadable. But they do
use brilliantly the landscape of contemporary eighteenth-century
agriculture to demonstrate how the natural bounty of America
and the availability of vast expanses of farmland molded the European religious and political heritage into something far more dynamic—something never before seen or even imagined.

Crèvecoeur was a materialist—where people live, what they do,
how they work, determines how they think and who they are. He
believed that the farmland of North America was everything, and
its rich abundance critical to fashioning a new culture. Crèvecoeur's American man, then, is purportedly novel and he is surely
different from any in Europe, because he has room and resources
to be freely exploited. The American is wholly an untraditional

creature whose successful existence has proved that free and "insignificant" men fleeing Europe can create a novel culture from an unforgiving nature, as the formerly despised become culture's guardians on the frontier. This "new" man is of course a curmudgeon who would be very hard to deprive of his newfound liberty, who would only with difficulty be coerced or uprooted—and he would not be fooled by the trend and jargon of town. He is as rough and unromantic among his urban peers as he is in his mute fields—in other words, a hard-nosed, no-nonsense American.

So Crèvecoeur wrote his *Letters from an American Farmer* in the belief that the emergence of yeomen and free landowners in America meant the genesis of a new egalitarian American culture. Muscular labor, now autonomous and in service of the individual, would create a self-confident, viable, and pragmatic novel citizen—in place of the passive serf and ignorant day laborer of past inegalitarian regimes of the European monarchies. Yet his new farmer-citizen was to be at odds also with the trader and near savage who leaves nothing in his wake, who was made brutish by North America's wild rather than taming it. Crèvecoeur's American agriculturists alone, who had created cultural order out of natural chaos—homesteads, cultivated fields, bridges, small towns—had hit upon that rare middle ground: freeholding yeomen neither rich nor poor, wild nor pampered, brutes nor sophisticates, day laborers nor absentee lords. American democrats were not to be coffee-house intellectuals nor a volatile mob eager for someone else's property.

Crèvecoeur's powers of abstract observation and analysis derived from his own unique background. He was classically trained at a Jesuit college in France—Latin remains untranslated frequently in the text. He traveled widely, holding a variety of jobs, and for at least seven years he was married and raised three children on his farm, until the tumult of the Revolutionary War forced him to flee America. Crèvecoeur farmed for less than a decade before his return to Europe, where he entered diplomatic service and became a literary figure in his own right in revolutionary France. His *Letters* were intended to put the experience of agrarianism in a wider historical context to a mostly literary and urban European audience, confused over and fascinated with democratic revolution in their midst. Though he was a genuine farmer for almost a decade, agriculture was but a parenthesis in his life, which was ironically spent largely in Europe writing about farming in America. His *Letters*, then—as generations of critics have pointed out—suffer from the paradox of an ex-farmer writing about what he will not or cannot any longer do. As a farmer turned professor as well, I can empathize.

Still, Crèvecoeur's letters were an immediate success among contemporaries for two reasons: the largely European audience was curious about the creation of this new social paradigm in America, and they wanted to know the natural esoterica of a frontier and rural lifestyle pretty much unknown in Europe. The ostensibly fictional account is actually a firsthand look at life in rural New England, and details the creation and management of a working farm. To European contemporaries, its collection of Indians, slaves, war, pestilence, and near starvation on the frontier reads like a traveler's fantasy of a far-off savage paradise. Indeed, antiquarians today often mine the letters for the preservation of firsthand data on the everyday life of an America in creation.

But, again, the book's real interest, past and present, arose from its literary exploration of a more important topic: What is an American, and was he really so "new"? What is the relationship between the cultivated landscape of America and the nature of its citizenry? What has American agrarianism done to improve upon the Western paradigm as practiced in Europe—and could the muscular and uncouth govern themselves without the guardianship of the academic and refined?

Over two centuries later, I think our citizens know less about American independent farming than did Crèvecoeur's Europeans two hundred years ago. This is a great tragedy, perhaps the tragedy of the last half century. They have completely forgotten the original relationship between farming and democracy, which he sought so carefully to explain. As a consequence, I suspect few Americans can define in the abstract what they were or who they are. Few of us work with our hands or become dirty from the soil, unless we're puttering in our gardens; those who do so for work more often wish that they did not. The labor of muscle, unless directed to the narcissistic obsession with the healthy body, is deemed unfortunate, whereas the work of the tongue is alone prized—that the two might be combined and thus become greater than either is ignored or forgotten. To Crèvecoeur the dichotomy of the effete intellectual and the brutish thug—so common in Europe—is solved by the emergence of the independent American farmer, who avoided through his autonomy, craft, and labor the pitfalls of both. I agree: to walk into a room of farmers is to see some of the most rough-looking yet highly intelligent citizens in America.

Two centuries later I have returned to Crèvecoeur's approach to update and conclude his thesis: just as he held that the formation of freeholding yeomen created the American republican spirit, so now the decline of family farming in our own generation is symptomatic of the demise of his notion of what an American was. Just as Crèvecoeur saw unlimited land, small towns, multi-ethnicism, the growth of a middle class, self-reliance, and a common culture as essential to the creation of Americanism and its democracy, I am afraid that I see that the decline of family farming, the end of the egalitarian principle of farm ownership, the growth of urbanism, the assurance of entitlement, and the disappearance of a rural middle class ensure the demise of Crèvecoeur's American. Just as Crèvecoeur worried about the landless serf, who was volatile and thus dangerous, so I worry about the middle-class American of the present consumer society, who has become a transient servant of his appetites. The crowd in Wal-Mart and at home before *Oprah,* and the very few who profit from it all, can be more alike than they think. As in Mr. Crèvecoeur's era, we are now once more creating a new American—and he is not merely "different," not merely a product of changed times and circumstances. We have lost our agrarian landscape and with it the insurance that there would be an autonomous, outspoken, and critical group of citizens eager to remind the rest of us of the current fads and follies of the day.

Crèvecoeur was not naive, and not a romantic entirely: freedom, egalitarianism, and democracy were possible because man in America had little leisure, less affluence, and found success or failure largely in his own efforts. Surfeit for the human species was as great a danger as poverty, sloth the more terrible peril than exhaustion. Education and contemplation without action—the near-religious faith of today's intellectual class—meant not impotence, but moral vacuity itself. It was not merely democracy that was important, but the type of people who created democracy.

To Crèvecoeur, like Aristotle before, man was tame only to the degree he was occupied, independent only as long as he owned property. Only through agriculture was the citizen in constant observation of how terrible loomed the animal and human world about him: man realizes the dangers of his own natural savagery only through his attempt at physical mastery of the world about him. Many men and women who undergo this experience provide a check on those who do not. Such farmers question authority and yet follow the law; they are suspicious of the faddishly nontraditional, yet remain highly eccentric themselves; they vote and work for civic projects and group cohesion, and yet tend to be happiest when left alone, these who historically have been democracy's greatest supporters by not quite being convinced of the ultimate wisdom of democracy.

In contrast, Crèvecoeur's trappers and traders who live as natural men on the edge of the frontier are not romantic individualists, but more often beasts—without permanent residence, without responsibilities to others, without desire to clean and separate themselves from the foul world they must inhabit and have surrendered to. They and the refined urban merchant both dwell, then, in antithesis to the farmer, who both conquers and lives with nature, who practices both a solitary and a communal existence, who is and is not one with the government at large. From that personal, strife-filled experience of working the soil, the yeoman-citizen alone transfers his code of stewardship, reasoned exploitation, and independence to the wider society of his peers, this muscled reader of books, this hardened lover of beauty. That the balance and stability of agrarianism in itself explains the health of a culture now seems to us of the post–Industrial Age preposterous. But to Crèvecoeur that connection was self-evident to the point of being unquestioned.

In the great American debate over ecology and development, and the use and abuse of nature, we have forgotten the central role of agriculture, which is more than just to keep us alive one more day. Farming alone reminds us of the now lost balance between wilderness and pollution, and inculcates in our youth the thought that true erudition is not the mastery of the specialist's esoterica, but broad learning, checked and tried daily through the pragmatics of the arm and back. We need to be retaught that your idea is only as good as your will and ability to see it enacted. The more abstract, liberal, and utopian your cant, the more difficult it is to live what you profess. The farmer of a free society solved uniquely the age-old Western dilemma between reason and faith, the balance between the Enlightenment and Medieval minds by using his reason and intellect to husband and direct the mystical world of plants, even as he accepted the limits of reason by experiencing every day a process that was ultimately unfathomable. The land taught us all that, and so was the nursery, not merely the breadbasket, of our nation.

We are not starving in this country. I have no worries about our food supply under corporate conglomerations to come. But we are parched and hungry in our quandary over how to be the good citizen—who the Greeks, the logical forefathers of modern democracy, said was ultimately the real, the only harvest of the soil. Because I know that one in one hundred Americans can now feed his peers, and that this one percent of the population is now almost entirely part of a corporate entity, I am worried where the country is going to find the spiritual censor and critic it needs, who is neither university pundit nor wild-eyed fanatic. I sense that we cannot afford to lose too many more like those now soon to exit this world, who are our last link with the founding fathers of our political and spiritual past.

Our new American is responsible for little property, other than his mortgaged house and car; his neighbors and friends—indeed, his very community—are more likely ephemeral than traditional and rooted. Although not an aristocrat, he is esteemed by his peers to the degree he is polished and secure, avoided once he is at odds with comfortable consensus. He depends on someone else for everything from his food to his safety. Lapses in his language and manners can end his livelihood; obsequiousness, rather than independence, is more likely to feed his family. The direction of the wind, the phases of the moon, and the dew point are as unnatural to the new American as his keyboard and cell phone are organic. He is angry today, content tomorrow, reactionary over this, liberal because of that—his entire ideology no ideology at all other than the expectation of material surfeit and liberty to enjoy his gains as he sees fit. Those of all stripes who promise him ever more bounty without responsibility and sacrifice he alone supports. Yes, America is more democratic and free, and perhaps a kinder and gentler nation than in the past; but political and economic advance came at a god-awful price. For a time we have become more humane collectively and in the abstract, but somehow far worse individually and in person.

II.

We American agrarians of the latter twentieth century fought a war for land that we did not even know we were in. Yet apparently we have lost it nonetheless. Family farmers as a species were mostly unknown fatalities in the new wave and final manifestation of market capitalism and entitlement democracy, the final stage of Western culture that is beyond good and evil. Ever more unchecked democracy and capitalism—because they alone succeed in achieving what they are designed for, and since there is no alternative to either—are now nearly global. In the next century both practices will ensure to the billions of the world material prosperity, entertainment, and leisure undreamed of by any generation in the planet's history. Surely billions will prosper as princes where millions once lived as the dispossessed in squalor, disease, and filth. Even the exploiters of capital cannot siphon the sheer abundance of lucre from the mob.

Yet this remarkable success has brought us to the end of history as we have known it. The age-old Platonic antithesis between what we can do and what we should do has been settled in favor of the former. There is no political, no religious, no cultural idea left that stands in the way of bringing more things to more people at any cost, of dismantling every cultural, religious, and social impediment to self-expression and indulgence.

Family farming, ancient and deemed inefficient, is gone. In its place we have, I suppose, modernism and postmodernism—currently the self-acclaimed cultural censor of the affluent society—offering cynicism and nihilism as smug cover for joining in the fray. I have never met a postmodernist who was poor, worked with his hands, or lived the nihilism he advocated for others.

In the absence of an agrarian creed, no intellectual has stepped forward to craft a higher culture for the people beyond materialism and consumerism. No abstract thinker dares to advocate the love of soil, a legacy of hard work, loyalty to family and town, country—even fealty to a common culture. No one suggests an erudition that is harmonious with, rather than antithetical to, muscular labor. These are the glues that hold—and should hold—a people together, that make their day-to-day drudgery mean more than the gratification of desire. Say that, and one would be dubbed a crank, misfit, and worse—corny, naive, and silly for sure. And why not? Everything that we hold dear—our mass entertainment and advertising, cars, leisure, music, material wealth, easy jet transportation, health, and consumer democracy with its moral relativism and cheap caring—are ours precisely and only because we have evolved away from the agrarian ideal and a vibrant countryside.

The end of family farming gave us more food—you must confess it, agrarian romantics—more time, more money, and less shame. Indeed, maybe even more equality as well.

At the millennium, time and space as the age-old brakes to human appetites have nearly given way as more material things and psychic leisure are now the promised entitlements to everyone on earth. In the end, the assurance of unlimited goods and services, and the convenience and freedom to enjoy them to surfeit, have proved more than a match for Marxism and autocracy alike. They have overcome Enlightenment thinker and starry-eyed Romantic both, humanism and religion together, the geriatric apparatchiks on the May Day dais, even old, lecherous Mao himself. Our new age is akin to the era of the so-called Five Good Emperors at Rome, the period between A.D. 98 and 180, whose monotony and materialism Edward Gibbon called the most tranquil period of human existence. Ours now is. "No other way of life remains," wrote the contemporary Greek toady Aelius Aristides of a past epoch:

> There is one pattern of society, embracing all. . . . Were there ever so many cities, inland and maritime? Were they ever so thoroughly modernized. . . . Seashore and interior are filled with cities, some founded and others enlarged. . . . The whole world, as on a holiday, has changed its old costume . . . and gone in for finery and for all amusements without restraint. All other animosities between cities have ceased, but a single rivalry obsesses every one of them—to show off a maximum of elegance and luxury.

Not just yeomanry, but even race, language, custom, and locale are falling before the onslaught of instant communications, advertising, unfettered speech, and material dynamism—before the idea that leisure and escape from muscular labor are the agreed-on prize. For the first time in civilization, real material overabundance, and at least the veneer of egalitarianism it spawns, are upon us—the ten-dollar sneakers on the illegal alien look and feel hardly different from the two-hundred-dollar designer brands on the corporate lawyer; the tap water of the welfare mom can be as clean as that of the exploiting blueblood; the video brings entertainment—any entertainment—as quickly, cheaply, and often to the illiterate as to the opera buff. Class, gender, skin color, and ethnic divides all blur through the satiation of shared desire—we know it, so the more our demagogues scream for even more evenhanded entitlement, the more our elites maintain their offices by promising even fewer standards and distinctions to come. Ease of consumption unites us more than race, gender, and class divide us. In short, for the first time in the history of civilization, the true age of democracy is at hand, encompassing not only the ideal of political equality, but a real material kinship and shared consumption at last. There are no longer the age-old skeptics from the countryside to come into town and remind us it is all dross.

The agrarian life, which is neither materialist nor fair, is the most visible casualty of what we have become in this age of Pax Sumptuosa. And we all have on occasion become willing casualties in this Faustian trade-off. It is baffling still to see one's children emerge exhausted from a day's hoeing of vineyard weeds with enough energy left to head right for their video game consoles. We poor farmers do not understand the present because we believe in ethical restraint on the economy; yet at the same time, as American consumers, we too want and expect what this efficient and amoral economy has to offer.

For most of my early adult life I was called a failure for farming; now I am dubbed a success for having failed at farming. Thus I can offer to the reader some insight about the consequences of the cultural demise of agrarianism through my own inability to live an exclusively agrarian life: I can write well of what I do not like, because in some sense I have just about become exactly what I do not like.

The alternate Western, agrarian tradition of autarcheia, autonomy, localism, and shame—which was always at war with our urban genius for materialism, uniformity, and entitlement—now has more or less lost out. It has always lost out, just as the *polis* has always given way to the kingdom, republic to empire, culture to civilization. They put too much responsibility on us, these voluntary checks on acquisition and consumption, on efficiency and bounty itself. The middling agrarian, whose age-old role was to preserve society from the dominion of the gifted but brutal renegade— Plato's solitary superman who would live by natural law alone— now gives way to the contemporary man of desires. The latter is full of reason of sorts but without spirit, and uses his knowledge mostly to seek complacency among his bounty. This contemporary clerk, teacher, salesman, or bureaucrat is everything the farmer is not: mobile, material, careful and timid; at peace with security, sameness, petty reputation, and complacency; glad for an endless existence of leisure and affluence without the interruption of strife or discord; nose always to the scent of cash and pleasure. He wants liberty, but too often liberty for indulgence alone, and then is surprised that when such commensurate license is extended to the less fortunate, they shoot and inject rather than show a taste for industry.

Agrarianism was such a brief interlude between savagery and decadence; it was such a hard teacher of the human condition.

Mr. Crèvecoeur, I fear we of the present age are now creating our own new American man, but he has nothing to do with either agriculture or the larger idea of an independent and empowered citizen. Rather, the new American will be largely a suburban creature to whom land will be nothing. He is surrendering his independence, his very will, in order to be liked, taken care of, and satisfied, this would-be man of action who idealizes a savage nature but only at a very safe distance. Food, clothing, cars, all material goods are simply things to be had on the mall shelf, with not a worry where they came from or why and how they are there, with the assurance only that they will be there somehow. With little physical challenge left to Western man's existence, with the assurance that all things must and can be as he wishes, that nothing is really his fault, and convinced that nature is kind and will always be at his call, the new American—there has been none quite like him in the long history of civilization—at last will receive what he wanted.

Americans: your cropland will be rented out to broke serfs and thriving corporations alike. Land will be lived on by harried commuters. Land will be gardened and it will occasionally be used by clever and tireless entrepreneurs to supply fresh fruit to nearby urban markets. To the untrained eye, land that is not urbanized will have lost its cultural value even as it appears pretty much the same as farmland always has. But is it the same? Will it serve a larger purpose beyond the production of food? Will it provide an increasingly rare impediment to what we are becoming? I think not. There was food being grown on land surrounding the Greek city-state in the Classical Age; and on that same land, food was still being grown for the Greeks in the centuries after. But the way in which that farmland was used, owned, and settled did not remain the same. And so there was not always a city-state. You see, the Greeks believed that the land produced not simply food, but also the free and good citizen. Collective farming in the ancient Greek world sustained the Mycenaean palace; feudalism, the Dark Ages; corporate agriculture, Hellenistic despotism; agrarianism alone, the *polis*.

The old conception of an entire family—grandparents, parents, and children—living from nothing other than the fruits of their labor, raising, not surviving by selling, produce; passing on a successful livelihood to sons and granddaughters; conveying ideas of independence, shame, and skepticism; and criticizing both the bookish and weak, and the robust and ignorant, will disappear. Indeed, it already has. And so I wonder, was the agrarian tradition of Western culture, the sum total of millions of mostly unknown existences and personal tragedies, of lost crops and ruined lives, all for this? Was the agrarian character of Mr. Jefferson's America to evolve only to give us the abundance, convenience, and freedom that we might become what we are? Was that what family farming of Mr. Crèvecoeur's age was all for? Was Mr. Crèvecoeur's yeoman to lead us to what we are now at the millennium?

Other good souls still bravely resist. Their attempts to re-create rural farming communities, to share in neighborly agrarian enterprises, and to forge farm communalism are noble and needed enterprises. Yet something will bother us about many of them. We will in secret confess that they are a bit scholastic. They are without the challenge and disaster of the past. This alternate agriculture of the organic gardener and suburban homesteader will be contrived by those whose daily survival and capital are really found elsewhere, rather than in the spontaneous enterprises of working farmers. In the century to come, treatises on resource management, indeed on the beauty and serenity of farming itself, will proliferate in inverse proportion to the demise of family farms. Well-meaning but—dare I say?—intellectually disingenuous souls will praise and romanticize family farming but fear to say it is gone. Such a literature will be artificial, largely the product of those with access to off-farm capital, or who for the most part were not born into or are a part of working farms (and who write for an elite even more removed than they from the land). Bib overalls, straw hats, folk music, the hoe, scythe, and horse, rural and informal parlance—"howdy," "folks," and "you-all"—will be mere symbols of a conscious but less than genuine attempt to recapture what we have lost. *Et in Arcadia ego* and "The man with the hoe" will be the counterfeit agrarian ideal. Suburbanites will respond to it. The therapeutic culture will embrace it. But it will be a lie.

In the postagrarian era to come, we who were not part of the Classical Age will do all in our power to restore it—a doomed endeavor, whatever our noble intent. Many agrarian idealists, as they have professed to me, will seek solace in such pockets of vitality as the Amish—those much-praised Amish, who can withstand the tide and hold their way thanks only to a fiery and uncompromising God, and to a surrounding unagrarian society that indirectly subsidizes them. Yes, restorationists will look to the crazed Amish. They prove that the horse and plough, dinner at five, asleep at nine, are yet possible if one will just suffer enough. But in the end even the most diehard farming reformer will not wish to be as the Amish are—and they will not know how to be like the Amish without being the Amish.

Their praiseworthy experience will emulate, but not continue, the agrarian idea, which grew out of a centuries-long tradition of families tied to particular farms of about the same size. At the end of agrarianism—when as with autos or steel, there are but a score of megafarms—we will find the demise of real conservatism. When all dour populists are gone, we will see that the market is not so conservative in its excess, and the liberal not so tolerant in his utopian agenda for his peers. The second-most-bothersome Americans I have met are globalist profiteers, who justify every exploitation imaginable as the inevitable wages of their market as deity. Perhaps the most offensive are very serious and usually affluent left-wing utopians, who foam and grimace from their elite white enclaves as they explain how we must all be forced to do this and that, here and now, to save some rare amphibian, inert gas, anonymous arteries, someone's lungs, or inner-city child's dreams—or else!

With the loss of this country's agrarian and conservative profile also goes a tradition of using agrarian life to critique contemporary culture, a tradition of farming as moral touchstone of some 2,500 years' duration in the West, beginning with Hesiod, Xenophon, and Aristotle and ending with us. Agrarian wisdom—man using and fighting against nature to produce food that ensured his family stayed on the land and his community was safe—was never fair or nicely presented. Family farmers prefer to be at loggerheads with society, yet they are neither autocrats nor disillusioned Nietzschean demigods sneering at the growing mediocrity of the inferiors in their midsts.

As their doomed and near-extinct status illustrates, yeomen are rather different from the rest of us. These Ajaxian men and women oppose us but mean us no harm—they are more suicidal than homicidal. They bother us with their "judgments" and "absolutes" and "unnecessary" and "hurtful" assessments that derive from meeting and conquering real challenge. But they also bother us in order to save, not to destroy, us by giving a paradigm of a different, an older way that once was in all of us. They want us to slow down, not to implode, to find an equilibrium between brutality and delicacy, as they themselves have with their orchards and vines. They want us to try something out ourselves before advocating it for others.

It is very hard at this age to pump your own water, drain out your family's feces, grow food for others from the ground, live where your great-grandparents were born, be buried beside your sister and great-aunt—and know that right now should the electricity cease, the phone go dead, the battery vanish, you and your own could still eat and drink, and survive one more season—and as citizens, not recluses or survivalists. Such folk who are ready—perhaps even willing—to do just that each day, look at the world radically differently from the rest of us. They laugh at most of our politicians, television programs, movies, and universities, where all men and women are to lead clean, safe, happy, and long lives. Their perspective, for all its involvement with the universal needs of life, is the most distant, and their worldview the most basic. In a democratic America, agrarians even now are more akin to the *polis* Greeks—the architects of Western constitutional government—than they are to the people of Los Angeles.

I think nowhere are the consequences of the yeoman's demise more apparent than in the laboratory of the San Joaquin Valley of California. It comprises the planet's richest irrigated agricultural region, but one which, at the present rate of development, is about to become one of the most polluted, congested, ugly—and increasingly dangerous—regions on this continent. It is no exaggeration to confess that America is intent on ruining the richest irrigated farmland in the nation, which produces one-eighth of its food and once provided some cultural sanity in the general chaos of California.

Yet, without our Valley of small farmers, America is not starving—our nation's fruits and vegetables for a time will continue to be grown by conglomerates and when that is no longer possible augmented by importation from abroad. But thousands of acres in California are no longer growing the citizens of old, who in the store, on the school board, and at the local meeting bring the rest of us to our senses through their wisdom gained from a solitary, daily war against nature, society, and self. We sense that loss daily, not merely because of the greater abundance of pretty, accessible, and tasteless fruit, but through the presence of a society that is ostensibly kinder and more humane, but upon tasting is in fact soulless and without spirit.

Family farming is gone, yet democracy and Western civilization remain, the creations of agrarianism. We Americans can survive, thrive even, under the material conditions of the next century, but we will never be anything like what we were. The hardest task in America is now not to fall into defeatism—even if it means verging on idealism, corniness as well. So I offer these letters, both brutal and naive, in hopes that we might still learn from what we are losing.

Part One

MAN VERSUS NATURE

AT THE ABYSS

(For Those Who Might Yet Visit Central California)

*Here he beholds fair cities, substantial villages, extensive fields,
an immense country filled with decent houses, good roads, orchards, meadows, and bridges where an hundred years ago all was
wild, woody, and uncultivated! What a train of pleasant ideas
this fair spectacle must suggest; it is a prospect which must in-
spire a good citizen with the most heart-felt pleasure.*

—J. Hector St. John de Crèvecoeur,
Letters from an American Farmer

Farmers are in a dilemma. We proclaim that we don't like modernism, yet we rely on, indeed often enjoy, the technology of the
age. We are going broke, but the nation at large is affluent. We say
that we are crucial to the country, but America has found a way to
feed itself quite comfortably without the presence of the family
farmer; indeed, the rise of cheap food and corporate farms are
perhaps explicable in part because of America's very wealth and
serenity. We claim to be conservatives. But our conservatism is of
habit and practice, without sympathy for laissez-faire capitalism
that rewards a chosen few who do not appear to us in the flesh as
real men and women of action and daring. We are uneasy with
cities and with corporate America, not out of ideological prejudice, but simply because we feel their life is not ours. We, who em-

body the values that made our culture majestic and whose failure is symptomatic of what has made the country materially rich, appear at odds with everyone and have no idea where to go or what to do. So mostly, we farmers just dream of the past, lie about the future, or rail about the present. I am no exception.

I do not think I shall leave the San Joaquin Valley of California. My children are the sixth continuous generation to live in the same house. For 120 years, from malarial ponds to Wal-Mart two miles away, some member of this family has lived on this ground in one of these five ramshackle farmhouses. I guess that I will end here too, like all the others, one more in a row of cement slabs in the Selma cemetery. Courage, a friend tells me, requires me to grow up and leave, to get a better job elsewhere; cowardice, he said, is to stay put, possumlike, as the world goes on by. But at least my credentials as a San Joaquin Valley loyalist are unimpeachable, and thus my lament over its destruction is genuine.

Until just a few years ago I argued that our Valley was beautiful, not an ugly place as most believed, and so it deserved some aesthetic repute. At Santa Cruz and Stanford I told professors and fellow students (whose Valley horror stories comprised little more than car boilovers in Bakersfield or gas fill-ups on the 99 freeway in transit to Yosemite) that its nights were unusually warm, its sunsets far better than those on the ocean. Out of consideration, I omitted that it was also a region of praxis, not theory. Here men and women with their labor usually paid the consequences for the ideas of distant others. No one has ever disparaged their legacy on the grounds that this is a laid-back, complacent, or even contemplative province without material foundation. No one has ever claimed that Americans came here to sight-see, vacation, garden, ski, hike, or shop—came to do anything other than work and suffer. As an agrarian culture, our hard work permeated everywhere, encompassed everyone.

I used to see very few cobblestone streets, cappuccino bars, or neighborhood theaters in any of our small Valley towns. Instead, the Valley's attraction was always and only its bustle of hardworking agrarian peers, and the bounty that grew out of their endless and uncomplaining work. For all our Valley boosterism, let us confess that there is little tradition of great universities, renowned symphonies, impressive museums, or enclaves of artists between Fresno and Bakersfield. Let us, then, live or die with our strength, which is the wonderful soil, the monotonous climate, and the agricultural genius and good sense of our people, a folk like none other on this planet and whose toil created an oasis.

In short, the Valley, I summarized to outsiders, was irrigated trees and vines in a desert, framed with snow on the Sierra Nevada, the evening sun over the Coast Range, and some half-million self-absorbed and autonomous immigrants on forty-acre farms in between—a relatively unknown but prolific food basket. Its buildings—the Fresno water tower, the imposing domed county courthouse, the original Sun-Maid raisin plant—were massive stone and cement behemoths, absolutely utilitarian architectural masterpieces. Only with difficulty did we dynamite that Ree-bar and concrete into rubble, or simply abandon them to the birds and rodents. Farmers of this Valley were, as a rule, strong builders—a real handicap when the function of buildings must change even if their form cannot.

When in Greece, I pointed out that Boeotia was like farmland outside Visalia, the Attic hills not unlike those right above Fresno. Citrus-raising hamlets of the Peloponnese and the Argolid resemble Exeter, Orosi, and Porterville. Did not the Valley share the Hellenic latitude, is not its climate similarly Mediterranean? Did not the most audacious Greeks, Italians, Spaniards, Basques, and Armenians as well, come to Fresno for its southern European seasons? When I returned from one visit overseas, I even planted a bay tree (*Laurus nobilis*), and it grew here next to the vines every bit as well—no, better—than those I saw in the Pindus Mountains, as does in general the Mediterranean triad of grape, olive, and grain.

The farms of this Valley, then, often match those of southern Europe, ancient and modern, and gain distinction for that stately resemblance. The farmers here are also far more ingenious, far harder-working than any I ever saw abroad. When the railroad opened this vast plain to homesteaders, for once its advance men were not lying when they said it was fertile. Even their tawdry hucksterism did not exaggerate our soil and climate. We were not Montana or the Dakotas, whose icy sterility made duped homesteaders weep. Did not Steinbeck and Saroyan see some natural and human beauty here on the canvas for even their bleakest narratives? And was not such a Mediterranean climate historically the laboratory for democracy, for agrarian timocrats who met outdoors to argue, shout, and vote? In short, I thought Greece had come alive again in the central San Joaquin—not a democratic Athenian Greece, but a tough Theban Hellenism of Hesiods and Plutarchs.

But let me be more frank about what we have created here in central California. Most of you readers would simply not like to move to this wonderful soil and climate. If you were stuck here, like most in the Valley of today, you'd say, "I'm leaving when the kids are grown" or "We're out of here when the transfer comes in," or, "If it was a choice between Tucson, Los Angeles, or Fresno, you'd have come here too." In its own way, the Valley is as unattractive as it was before man in any numbers arrived. Its Golden Age was therefore brief, no more than the beautiful century between 1870 and 1970, when gravity-fed irrigation in hand-dug ditches from the Sierra first turned a weed-infested desert into an oasis of small tree and vine farms and their quiet satellite communities.

Is too much man as bad as too little? Farms that arose out of wild scrub now, in turn, fall before man's latest incarnation of cement and petroleum. The farmers? They are all in the rest home, dead, or tend bar, teach school, and patch tires to tide themselves over for one more money-losing peach or raisin harvest. All that really remains is the romantic ideal that we in this Valley might still appear, and be, as our forebears were—without ever living as they did.

After that brief hundred years of agrarianism, developers and industrialists picked up the scent of the San Joaquin's natural and human attractions, and so moved factories here and built housing tracts by the thousands. Their ongoing bustle is now our shared legacy. Shortly it will become our joint agrarian epitaph. I fear that we are here in the Greeks' age of iron, where the brutality of life is not softened, but rather enhanced, by a landscape that is becoming wholly suburban where it should not be so. What killed us was our success, not our failure, our fertility, not our barrenness, our beauty, not our ugliness. The farmer laments, it must be confessed, not just his own demise but, to his shame, the material success of others as well.

On a local level, we are confirming the last chapter of agrarian history. Family farming, sustainable agriculture, communities of yeomen peers, and their accompanying cargo of free speech, consensual government, and natural inquiry unfettered by government or religion, are rare historical occurrences. Like the comet in its centuries-long parabolic orbit, these cultures appear only for a while and do so with a radiant burst—the Greek city-state, Republican Rome, eighteenth- and nineteenth-century Europe and America, perhaps now the spark to be rediscovered again in rural China. And then these agrarian communities disappear, after providing for others the safety and material foundation for a slow, though thoroughly enjoyable, decline.

In contrast, palatial bureaucracies and vast estates are civilization's drab norm, where farming is food production as part of capital exploitation. An urbanized population with a surrounding dependent countryside is the more common paradigm—with its detritus of serfdom, peasantry, a wealthy overclass, and political autocracy. Thus as historians, we must acknowledge how rare our Valley century was, that the Mayas, Aztecs, Egyptians, Hittites, and Gauls were not like the *polis* Greeks, whose citizenry owned and was of the countryside. We must also appreciate, even romanticize, those singular hundred years when the brutal polarity of masters and serfs was but for a time abated. There once was a real middle class of spirit and circumstance, who lived either from the land or from those who farmed the land.

But very different people of the last two decades have come here to pave and subdivide because of the foundations the agrarians had established before. Would they not say, "You did a fine enough job here with what you had, with your picturesque little towns and farms; but now step aside and let us show you what can be done when you have real money and more than enough people to finish what was started."

Those elder yeomen now dead came to this Valley between 1860 and 1900—I mean the eastern side of the Valley with the water and near the Sierra—and planted and beautified. Like Lucretius' Roman *agricolae*, they "set shoots in branches and buried fresh cuttings in earth about their fields. They tried to grow first one thing, then another, on their beloved lands, and saw wild plants turn tame in the soil with coddling and gentle, coaxing care." Thirty years ago you could have agreed with Lucretius that here too were "lands everywhere marked with beauty, lined and adorned with apple trees; and fruitful orchards wall them about." Hard work on the land produced a stable citizenry and a beautiful countryside.

But not now. In this 1990s Valley we are becoming wealthier, freer, and more numerous. The air is at once foggy and smoggy in winter. Forget our lungs; even the vine leaves burn in the ozone. My grandfather in the last year of his life, 1976, was baffled at the scorched leaves of his vines. After brooding for days in late summer, checking the vineyard for insect damage, worried about sulfur burn, complaining about insufficient water in the ditch, he finally blurted out, "Why, I think it is the very air. It's too dirty now for those vines."

Thousands of us rise at dawn not to harrow or irrigate, but to cram Toyotas and Hondas onto the 99 freeway and pour north into Fresno, three lanes of bumper-to-bumper cars amid paved-over and abandoned vineyards. I too have joined that *bellum omnium contra omnes.* If in the old days we worried about the hundred or so each year infected with the Valley Fever, the fungus that got into the lungs and then brain of those newcomers, now the enemy is the atmosphere itself. The peril is not the trace toxic spore that floats up from the dirt. Rather, the pollens and particulates raised by agriculture combine with industry's ozone and monoxide to create a uniquely ugly haze. It gives our people perpetual allergies and asthma. No fungicide shot into the cranium can cure it. Our air is worse than our once dreaded indigenous fungal spores. Asthmatics of Fresno say their breathing improves after a week's stay in the Los Angeles basin.

We few romantic and autochthonous natives of the Valley are in a race that we cannot win, between land farmed and land paved. Our ancestors beat back the wild to give us these farms—we of a lesser breed were beaten back by new men to give them up. We battle the encroaching suburb; our grandfathers in their youth once fought hunger and typhoid—both of us have received four dollars a box for plums.

The developer and industrialist know that the end of unlimited exploitation is near and the battle now won. So they build and overlay at an even more frantic pace to grab the last air, water, and land from farmers who are left, before nature changes the rules of the game and we are all but another Los Angeles. When we are one megalopolis between the mountains and the 99 freeway, it will be rather difficult to find a peach orchard to bulldoze under, an aquifer to suck dry, or a farmer to buy out. (Then beware, you remaining pristine valleys elsewhere.) It will be a more difficult task for the developer and speculator to make a lot of money when the entire Valley corridor is cement, asphalt, and satellite dishes, with only a few skipped lots in the mess, just a tessera or two missing from this bad Roman mosaic. And we will wonder why we have become a different people.

Agriculture will be doomed here in fifty years, we are told by the more honest futurists, when the air, water, and housing sprawl of our Valley will finally turn traitor on us. Farms that survive will be rural atolls in an unending suburban sea. After all, in the next century we will have well over 15 million suburbanites between Sacramento and Bakersfield. In short, we who came of age in the generation between 1970 and the present, for all our energy, work, and suffering, for all those unknown and little-heralded sacrifices and tragedies up on the scaffold, on the roof, and in the trench, have almost ruined in little more than twenty-five years what it took a century to build—and we have called that subversion a great success. And if success means the greatest number of people in the history of this Valley well fed and housed, it surely is a triumph. Land that once fed millions now houses and employs—and still feeds—millions more.

Give our frenzied cohort its due. There was muscle and there was sweat involved in these last two decades to give us our malls, fast-food drive-throughs, and three-bedroom tract houses. There is courage and there is sacrifice among those whose sinews helped to ruin this Valley—unknown but noble pavers, cement workers, roofers, and carpenters, who died mostly sore and poor. Nothing in this state's history, not the original immigration from Mexico nor the trek of thousands from Oklahoma during the Dust Bowl, can rival the sheer rapidity and totality of what my generation did to the San Joaquin Valley's landscape. The adulteration of our soil, sky, and aquifer gave us bountiful housing, entertainment, and people.

What then have we lost? Irrigated valleys, whose rich loams stretch out on flat plains, where water in the mountains above is only thirty miles away and gravity-fed, where neither hurricane nor ice intrude, are not common conditions on this earth. Perhaps in Chile or South Africa, and maybe in western Asia Minor and southern Europe such oases are found. But they are still smaller there and less opulent, and rarely do their inhabitants have the freedom to realize the potential bounty of the natural treasure they have found.

The San Joaquin Valley of California is a natural slip, an enormous mistake of the gods, who rarely give farming man so much natural bounty, so much dormant power to sprout from the once countryside of scrub. Two hundred fifty crops now produce $13 billion worth of food every year. Its inhabitants too are historical oddities, refugees from every agricultural area in the world—China, Japan, Europe, India, Armenia—who, as Mr. Crèvecoeur saw, fled here under the aegis of liberty to enact without coercion their own particular philosophy of the growing of food. In human terms we are the world's receptacle of unappreciated farming genius, let loose in the finest combination of earth, water, air, and heat on the planet. What a fertile Valley we have been and shall not be again. They are now forgotten and neglected, those giants who had so much faith in themselves and so much courage before the terrible wages of their sacrifice. As Mr. Crèvecoeur said, "Is there no credit to be given to these first cultivators, who by their sweat, their toil, and their perseverance have come over a sea of three thousand miles to till a new soil?"

Why and how then did we do such things to this basin of greenery? I had always thought that our snow pack in the mountains to the east, the natural fertility of the soil, the extended and cloudless growing season, the presence of a capable and hard-working population, would confine urban growth. Such largess assuredly would convince even the most grasping not to ruin those precious ingredients of irrigated agrarianism that existed nowhere else in the world. Anyone knew you do not uproot a rare thirty-ton-to-the-acre peach orchard for one more Motel 6; anyone should have known that the local megacasino, sprung from a vineyard, now itself bankrupt and empty, was a bad trade-off in the use of land.

Cheap food, cheap housing, cheap transport can also give us cheap people. To drive into a once agrarian backwater like Selma now is to witness the result of this radical venture: streets that are not safe after dusk, millions of dollars in material goods purchased on credit by those who have no money and no idea what to make of their third cars, pagers, cell phones, and assorted sundries, a populace that is far better clothed, fed, and tended to—and more ignorant, dangerous, and unhappy—than four decades ago. We have now become Mr. Gibbon's "crowds without company, dissipation without pleasure."

My grandfather knew better. The third of six generations to live in this house, born here in 1890, he was not beaten down by the past century of wars and depression. Instead, typically so, he used to worry about other more insidious enemies to come after his demise. "Why, I tell you this is God's country, not a place like it in the world, and sometime someone's going to find out about it. God help you kids when they do." They did.

Our natural advantages turned out to serve even better those who would not farm—those who saw this Valley's future as a bustling metropolis of teeming urbanites, not a backwater agrarian refuge. These commercialists, inventive and audacious folk all, were not bothered by our reputation for rusticity and yokelism. They welcomed the absence of symphonies and ballets and museums of the first rank in Fresno and Bakersfield. As purveyors of inanimate capital, the clever investors rather liked our repute as the Theban dullards of California. They were not snobs at all. They were, to be honest, real democrats of old, who wished to people this Valley to surfeit and knew there were few to stop them. And few did.

Unlike us, these antiagrarians were not trusting with nose to the ground; they knew how to acquire what was once ours and expand upon what they took. They knew that few farmers would buy out their failed plants and new soon-to-be slums, to rip up concrete and replant orchards. Successful or otherwise, their work remains forever. Ours will not. Even in defeat and bankruptcy, the commercial developer wins the race to replace farmland. Rip up a vineyard, and farming ceases—for all time to come. Abandon an unneeded and overrun apartment complex, and a vacant parking lot, crumbling storage yard, or weed-infested asphalt—not an orchard—takes its place. In this lifetime I have never seen a fruit tree or vine replanted on deserted and unwanted commercial real estate, the soil once more to be set free from its veneer of asphalt. Our local Wal-Mart, or what is left of Wal-Mart, will be here in a hundred years; the farmland outside my front door will not be.

So the developers like—no, adore—this Valley, which they say was better for their purposes all along. It is easy to see why. First, our Valley is dry and hot. Thus it is as good a place to keep roads, trains, and airways open as it is to dry raisins. You can hammer, fabricate, and nail shingles almost every day of the year as well as prune or disk. No one is barricaded from work by the snow. Tornadoes do not destroy the box plant, just as they have not uprooted orchards. Earthquakes are but slight tremors beneath our Valley's cushion of loam, shaking neither studs nor barns. The cold does not put welders into depression or prompt one drink too many, as it has not killed vineyards or ruined peach trees. Our university growth institutes confirm to the potential developer that there are fifty more days here of working weather than at his present abode. As relish, they add that one acre of asphalt and aluminum can sprout more jobs than can ten of trees and vines. And it can—and often with less water too. They do not always lie. I have learned a rule in America: wherever family farms thrive, be it due to culture or climate, suburban development will follow them. Its corollary is just as valid: wherever corporate agriculture, industry, or urbanism is in full swing, new residents will stay away.

There is something wonderful for Americans in that moment of transition from farmland to city wasteland. Agriculture's evaporation gives a final brief moment of stability, greenery, and values that we all savor before we ruin it. It is savvy business for developers to have backyards border walnut orchards, to ensure vineyards are nearby the new tracts, to lure suburbanites with promises of peach trees and grapes but a mile from their door. Developers prefer not to develop next to other developers; and developers sometimes imagine that they might live in their own developments.

Weather is not the Valley's only attractive enticement for the forge of the modern-day Vulcan. Things are cheap here—food, fiber, and fuel. We grow every type of comestible, billions' worth and more each year. Our fruits and vegetables feed the East Coast. We are the world's richest cotton- and dairy-producing centers. Grain, hay, and rice are grown at higher per-acre ratios than anywhere in the world. And there is still oil near Coalinga and Bakersfield. In other words, the cost of doing business near Fresno—eating, burning gas, traveling to and fro—is for a while longer inexpensive. We supply our own, and ship the natural and man-made surplus to you. We are the last cheap place in California.

We do everything first rate, and that ensures we will soon do none of it, as each day acres of trees and vines give way to houses and roads—the water a little scarcer, the air a little dirtier, the people a little more numerous and unhappy. It makes sense for the Los Angeles or San Francisco captain of industry to bring down his plant or business to where his employees can at least afford to live on his cheaper wages for a while longer. VINEYARD KNOLLS. THREE-BEDROOM, TWO-BATH, FENCED AND LANDSCAPED, STARTING IN THE LOW 70S, HUD FINANCING AVAILABLE, the signs now proclaim in flat uprooted vineyards throughout the Valley. Next year or so, 170 such homes are to be zoned to be planted a mile from this farmhouse, dwellings that are clean, cheap, well built—and no doubt already sold. They and the Food-4-Less and Payless down the road are as egalitarian as anything that came out of old Athens.

No wise man in California has ever said, "Pave over no more farmland and your cities shall once again thrive." Only then might our endless cycle of build, abandon, and move on cease. It really would work, this idea of a greenbelt around a vertical city with population and growth curtailed. Downtown real estate would regain value; vacant lots in town would be prized; police, fire, and other municipal services would no longer be stretched; and the citizenry would deal with, rather than flee from, the nightmare of the American city—if we could but pave over no more farmland.

But that tonic is, again, worse than the malady, for it is not altogether free and equal. "Who is to say I can't have a yard?" "Who gets to live there, while I'm over here?"

The state has located most of its new prisons here. There is a rationale to those moves as well: we have many unemployed but conservative workers, who, when given the chance, do guard well (if not at times a tad too brutally); we have flat, empty land that is congenial to walls, wire, and cement; and we have many criminals who can make the trip short from crime to court to cell—and we are largely free of the Sierra Club, the Environmental Defense Fund, and Nature Conservancy. Those northerners still ask bothersome questions before concrete is poured and do not value jobs that accrue from the underbelly of modern urban life. We, who a century ago dammed the Sierra streams without a blink to beautify the Valley, can hardly now hesitate to bulldoze vineyards. Introspection and vacillation are just not in us—even if we are now demolishing rather than creating agrarianism. I suppose you on the Left will say that we century-long reckless exploiters of the original wild are getting what we deserve. And you would be right.

Each little farming community of this eastern corridor of the Valley now has an "Industrial Park" and a Chamber of Commerce recruitment "team" whose job is to snag more suburban "light" industry ("expanding the tax base") and retailing ("more service sector jobs") that will ruin their downtowns, deracinate their blinkered society, cloud their air, and pave over their orchards. Years after their success, for some reason those more honest hucksters at the brink of the abyss are known to confess—in moments of religiosity, or morose at their retirement dinners, or perhaps deranged and cancer-ridden at their final interviews to the local weekly: "I guess we didn't quite realize that we more or less made a whole new town out there by the freeway, and an uglier one than the one we already had at that." Then these Cassandras die. Only the most shameless of the petty millionaires, like one local baron still in his prime at fifty, boasts of their destroyed vineyards and newly arisen edge cities.

Ironic it is that the only real expanse of uninterrupted greenery left in central California comprises the cotton, tomato, and rice fields of the West Side of the San Joaquin, the corporate side of the Valley between the Coast Range mountains and the 99 freeway. There the arid land ranges from 5,000- to 30,000-acre blocks—some agribusiness conglomerates now control 50,000 acres and more. The company towns have no middle classes; their massive sheds are empty of their land barons, who live as absentees in the gated estates of Fresno; the growing salinization, leached selenium, and general absence of anything planted taller than fodder, all ensure that no one wishes to move there.

Some migratory birds from their wastewater artificial ponds have one wing. Others sometimes flop about with three feet. Some are found with two bills. Imported irrigated water percolates no more than ten to twenty feet below the surface and then becomes salty and foul. Cotton defoliant in autumn sickens even the car-bound on the highway. And food grows there still in abundance. To be fair to those magnates, you would not live there either, even if you owned it all. The earlier Spanish and American annalists told us not to plant there, told us it was a vast salt flat beneath the surface that blocked drainage, told us there was no water to be had. They were right. Yet those "farms" thrive when we cannot. The federal government brought ditches from the north to pump the water in, and now it must bring them a sewer to the sea to drain the miasma out.

The corporate West Side survives pristine due to its distortion and deformity of agriculture; we, the far better culture to the east thirty miles away, perish for the beauty and humanity of our trees and vines. There is a reason, after all, why few Greeks wanted to live in the expansive and royal pastures of horse-rearing Thessaly, why most instead preferred the agrarian patchwork of Argos and Attica. My point? The destruction of agrarianism, which has a certain logic in economic terms—cheap, instantaneous food on demand from great distances requires vertical integration and enormous capital— makes no sense at all in its aesthetic or cultural ramifications. Our genius in America can provide the entire planet with all the food, goods, safety, schlock culture, and hygiene it wants, with everything except how to employ that bounty so that we are better, not worse, souls for our success.

Pindar said "water is best." It is. One can judge a society's moral character by the purity of its subterranean water. Ours, once crystal-pure and pooled to surfeit in the aquifer just beneath our feet, now is tainted. But it is still sought after and so increasingly scarce. We want what is disappearing below, chemical-laden or not. No wonder the corporate farms are selling their federally subsidized allotments to Los Angeles. Over there the aquifer thirty miles from the Sierra drops off to several hundred feet below the surface. Even here near the mountains, at its shallowest, it has begun now to fall a few feet every year. Our answer? Drill deeper wells, of course. I am saving up money right now for yet another. A 1,500-gallon-a-minute well is just what I need to water vines faster than those in town can suck out the pool beneath. On this farm we have lowered all the pumps twenty feet in the last ten years, pumps whose subterranean bowls were frozen at a steady forty-five feet seventy years ago. Our droughts of every six or seven years now do more than make things a little dry. They lower the water table another ten feet at a crack—forever. Even the wettest year of the century does not replenish what is lost.

The hydroelectric power from the Sierra dams nearby has long since failed to sate our Valley's electrical appetite. Our planners, once thwarted in their utopian vision of two or three big nuclear reactors, still talk grandly of coal plants (with new "scrubbers," of course). The fuel is to be mined in Utah and sent rolling in here each morning, for miles a constant roar of a hundred train cars clanking in and unloading at a time. (The mountains on each side of this very level Valley create good acoustics.) People on the coast of the more maritime and elite cultures, where ocean breezes blow the fossil smoke eastward to us, who play Athens to our Thebes, still block power plant construction. We of this basin, who live where the offal goes up a few feet and stays, welcome them in: without additional electrical power, there can be no more people, only raisins and plums. In short, we in the Valley have come full circle. If in 1850 our Valley was untapped and useless ground for farming, so too it will be again by 2050, when the room, water, and air are exhausted.

There are a few places still—one small tract lies on our own farm—that have never been cultivated and thus serve as museum pieces of the natural landscape before the nineteenth-century arrival of the exploiting man. Such ground is, of course, ugly. Full of weeds and wild willow, by May it is no more than scorched earth, its natural potential somnolent, sleeping but for the coming of man. That pristine lot is—forgive me—a natural trash heap that the local ecologists describe as "a rare pristine ecosystem full of indigenous grasses and insects." That weedy blot does offer a radical contrast to the verdant trees and vines that surround—the former gives no life to man, the latter bounty to those who would work. Despite what the university pundit says, that quarter acre of unspoiled waste is nature's wild fraternal twin to man's nearby Taco Bell and Auto Mall. Both are ugly in their own unique way; both are part of no real *cultura*.

A century later, our latest local contribution to civilization's bareness, every bit a match for that land useless without man, is down the road next to the interstate, an edge city growing along the 99 freeway one mile and a half from this farm. There man is in

abundance. There in my own small town—no better or worse
than the others—we have torn up vineyards and now have planted
the following crops: Wal-Mart, Burger King, Food-4-Less,
Baskin-Robbins, Cinema 6, Denny's, Wendy's, Payless, Ander-
son's Pea Soup, the Holiday Inn, McDonald's, Carl's Jr., Taco
Bell, four gas stations, three shopping centers, two videotape
stores, and a car wash—an overnight emporiopolis, a modern car-
avan stop to serve 100,000 drivers, twenty-four hours a day. Its
neon, smoke, and noise are an enticing Antioch or Tyre, a reminder, like our section of scrub, that both man and nature in
their extreme can be altogether ugly creations without any of the
beauty of the farm in between. We should recall Mr. Crèvecoeur
about growing man, the plant:

> Men are like plants; the goodness and flavour of the fruit
> proceeds from the peculiar soil and exposition in which
> they grow. We are nothing but what we derive from the air
> we breathe, the climate we inhabit, the government we obey,
> the system of religion we profess, and the nature of our em-
> ployment.

Old Columella said of the Roman mob, "We quit the sickle and
the plough and crept within the city walls, and we ply our hands
in the circuses and theaters rather than in grain fields and vineyards," with the result that "our young men are so flabby, so enervated that death seems likely to make no change in them." I saw a
congregation of Columella's folk, our modern "disadvantaged"—
pants drooping from their posteriors, precious metal in their ears,
$200 worth of jackets, sneakers, jerseys, and backward caps, the
raised middle finger and obscenities their salutations—just this
morning at the neon agora near the interstate. These young men
are free and they have leisure; and by world standards these young

Americans—objects of so much liberal solicitude—are not materially poor. The bellies of the most spiritually impoverished humans on the planet are not swollen from want; their blood is not wracked by bacteria, their clothes are not tattered and worn.

In contrast, anyone who saw this Valley, its orchards, dams, vineyards, and small towns, might wonder what type of people created such things out of dirt. Who could craft such beauty, feed so many, work so unceasingly in but a century? They were builders and artists, those who gridded off this scrub expanse with 20s, 40s, and 80s, a patchwork of vine and tree, cut across by road, canal, and farmhouse. Even the positive that we do here now is but elaboration on the labor of our forebears. But more often it is not even that, but simply cynicism, nihilism, guilt, and immobility. When the Valley's farms are gone, as they must be, I worry: without the challenge to tame nature, where will the citizen of this current society learn of his true potential, and where will be the physical space— away from town and yet not the empty wild—to refresh his soul? Where will a man learn that if he just works, he can still plant, still grow, and need not feel impotent before nature or man.

An empty San Joaquin Valley desert was ugly. A congested sprawl of 20 million will be worse. A checkerboard of a few hundred thousand green farms alone was a century of paradise now lost. That is all agriculture is—the fragile balance between the neglect and the ruin of nature. Farming is the mean where man can cultivate the wild and neither destroy it nor be destroyed by it, the rare equilibrium between the work of the mind and labor of the back, the tough community where the autonomous and pragmatic create and then follow the law.

For a brief age, we harnessed nature here, but did not desecrate it. Through the muscle and sweat of that struggle, we were taught to value what we have and not to wish for a great deal more. We now see that the Valley, which grew out of weeds, reached a brief perfection, and then reverted to the wild from which it came, man now doing far more evil through his frenzy than through his earlier neglect. That struggle to create the Valley taught us humility and measure: when we sweat to grow plums, we appreciate in turn that our showers are warm, our homes cooled. It is so hard each day to rid a vineyard of weeds, and so we felt victorious that our homes were safe and our police lawful. We, who shovel furrows and are scorched so that our vines can drink, are amazed that thousands can at last sip chilled water through their refrigerator doors. Since we see rodents gnawing our shoots, viruses sickening our fruits, and frost turning our vineyards black, we arise each morning convinced that nature's brood would like to do the very same to us—and will, should we cease the labor of the back. Farmers learned and passed on all that knowledge from the struggle to create this oasis, but their university in this Valley is now beneath our feet.

THE UNSEEN ENEMIES OF AGRICULTURE

(Wisdom from a Relentless and Unknown War)

I should have never done, were I to recount to you all the inconveniences and accidents which the grains of our fields, the trees of our orchards, as well as those of the woods, are exposed to. If bountiful Nature is kind to us on the one hand, on the other she wills that we shall purchase her kindness not only with sweats and labour but with vigilance and care. These calamities remind us of our precipitous situation. . . . One species of evil is balanced by another; thus the fury of one element is repressed by the power of the other. In the midst of this great, this astonishing equipoise, Man struggles and lives.

—J. Hector St. John de Crèvecoeur,
Sketches of Eighteenth-Century America

Early one May, I walked through our Castlebright apricot orchard to investigate the damage of an unseasonable spring. I paused beneath the eleventh tree of the first row. From there any observer could sense that on all 484 trees in that acre there were not more than a few dozen apricots—maybe sixty apricots instead of a hun-

dred tons, $10 worth instead of $40,000. It had rained during
bloom. What few flowers that had pollinated soon rotted off
anyway.

Farther down the row, through the barren trees, my neighbor's
new tract house next caught my eye. A former policeman had won
a generous and assured disability pension and so had moved to
the country. Immediately he bulldozed our communal road—
deeded easement or not—and began encroaching on common
ground, now complaining about our eighty-year-old power lines,
now uneasy about the vague demarcation of legal borders. In general, he was eager to upset the century-long tranquility of his predecessors that was characteristic of the eastern border of our
farm. When I reached the end of the apricot row, his garish sign—
FOR SALE, BY OWNER, INSPECTION BY RESERVATION ONLY, now
faded—was still there after a year. They come and they go, my
grandfather used to say.

I sat for a while on the border of the orchard and looked down
its lines and diagonals. Nearby to the north was another urban
refugee on a two-acre ranchero. He was—if the same person still
lived in the house—a truck driver who went mysteriously each
week or so to Mexico. Ray or Javier something. I never knew,
never wished to learn his name. No farmer himself, he came to the
open spaces here to park his rig and to raise pit bulls in peace—
Satanic-looking pups born near the equator and gathered on his
southern travels. Years earlier one of his brood had almost mauled
my then four-year-old son and me while we were irrigating the
apricots.

Now I moved away out of the barren apricot trees, and on this
May day caught a glimpse of the year-old Holiday Inn a mile or so
distant on the horizon. Nearby, the rotating propeller blade of the
fake windmill (a local travel lodge/restaurant/amusement com-

plex) came in and out of view. Over there the food and sundries were cheap, accessible, and clean. All these new nearby establishments on the freeway, even if for the wrong reasons, nevertheless did their part to help make this present democracy work.

But I knew from the Greeks that farmers are by nature not, and have never been, radical democrats. They would rather have had the vineyard than have it ploughed under for Wal-Mart, even if that meant that the masses would have to do with less. Classical Athens was not their city. They were folk of an earlier, timocratic *polis* of landowning hoplites, a Thebes, Argos, archaic Athens, or Elis. Agrarians' great strength has always been their autonomy, their distrust of materialism, and the chauvinism of those with both feet on ancestral ground.

Our flaw? We are not altogether tolerant of either the serf or his master, and so for rather ancient reasons we distrust those who both shop at and profit from things such as Taco Bell and Payless. Even at the edge of the orchard I could not count all the franchises by the freeway. So for relief I began to look for the mountains. Although I was standing with an uninterrupted view of the Sierra only thirty miles away—the avenue that dissects our farm and is aligned perfectly with Mt. Whitney is called Mountain View—the Valley smog made all beyond the apricots murky. With a view like that (one can both imagine away the smog and remember one's childhood), with snow like that only a day's hike from this farm, no wonder my great-grandparents and their parents 130 years ago came to homestead this spot.

By then, it had been one morning enough of looking at commercial democracy, of pondering outlaw dogs and ersatz windmills and the new edge city on the freeway. I walked back west toward home among the safety and familiarity of our land. Our high-school crony, Ramon Orosco, once the feared bouncer at the local pool hall, now fiftyish, obese, and diabetic, was spraying cryolite mixed with liquid sulfur on the vines. I thought I sniffed miticide mixed in too. "You working or still playing with the books," he screamed over the screeching agitator, as he made the turn into the next vine row. And then he was gone. Swallowed up, he was in a cloud of white cryolite mist. "At least it is mined from the ground," I thought. My cousin sort of waved, as he piled up dead vines from the hundred-year-old vineyard we had just bulldozed. Robert and Eddie Aragon were repairing fifty-year-old broken wooden ladders and looking to see if I was watching their job. I wasn't. My twin brother was in town engaged with the bank, and I put that particular thought out of my mind immediately.

I made my way home through the adjoining orchard of peaches. But I started to become further annoyed when I spotted in this grove gopher mounds, crows in the canopy above, sap oozing from the gummosis of the trunk at my left, curl fungus on the peach leaves, even soiled tissue paper and feces from the undocumented workers of a neighboring truck farmer, Cruz Franco, who was either too cheap, too mean, or too stupid to provide a bathroom for his workers. Enough battles for a week, all these. At that point, as I picked up my pace and finally entered the house yard, I began to wonder just who were all these enemies of agriculture. And what wisdom arises from the battle against them?

By noon I was at the desk and had scribbled out a typology of all these morning foes and more—I recalled the pit bulls barking, the two foul neighbors eyeing me suspiciously (in our orchard!), the rig's toxic mist rising about Ramon, and the dead vines soon to go up in flames. Oh, and the feces and oozing sap too. In minutes, playing with books or not, I had catalogued quite enough worries to write a proper letter of explication and warning to you, reader. I also checked the *Letters* of Mr. Crèvecoeur, only to learn that he too was aware somewhat haphazardly of the foes' presence. Two hundred years ago he had called them "the enemies of farming."

While I have lived on this farm most of my forty-five years, it has now been only a few hours since my knowledge of the war against the farmer's adversaries has become clear to me in any systematic way. So I record this rogues' gallery as it seems to me after my morning walk. If the enemies of agriculture will always be those that appeared to me beneath our orchards and in our vineyard in May, I must present them for you now. Know their dangers in advance, reader; all you, who have no farm to teach you, should realize that these adversaries are among you too right now.

The appearance of the soil, the vine, and the tree seems to remain the same each generation on a farm. The enemies of agriculture and man's counterresponse to them, I know, do not. Ostensibly, with the rise of technology the American farmer can gain the upper hand over those who would take his fruits; ostensibly—if the fight were solely against the unchanging bug or rodent of the ages. In reality, the farmer's more clever adversaries themselves mutate. Old enemies with new faces ensure that the war against the food producer goes on with stroke and counterstroke as it always has, though now in more complex and difficult terrain. Otherwise, old Theophrastus of Lesbos would have saved us long ago, with his brilliant 2,400-year-old treatises of warning about those who suck, eat, rot, and tear the life out of the farmer.

Agriculture, I think, will always be war. At the conflict's most dramatic, during an unseasonable storm or foreclosure warning, the agrarian fight becomes real bloodletting, a brutal, horrific, yet sometimes heroic experience. But most conflict and disaster on the farm is not so dramatic. Too often it is a struggle of a different kind—Hesiod's less romantic, more mundane sort that kills more slowly: family squabbles, poor prices, bad decisions, missed

sprays, wrong fertilizers, the rural counterpart to suburban deso-
lation. I confess farming is mostly a prosaic, dirty, and petty backwater skirmish against a nature that would insidiously deny the
farmer his power to raise up food from the ground. Often the hy-
dra's heads are not even grotesque.

But because farming involves nature and so remains always the
most basic elemental conflict, both epic and guerrilla, there are clear
enemies of agriculture, sensational and ordinary alike. From their
destruction accrues a larger, precious knowledge for the rest of us
about whom to fear. This wisdom is not to be found elsewhere. It is
in no great book that I have read, on the lips of no professor,
prophet, or French savant. It is not to be found with them and their
work, because unlike most other occupations in the history of civilization, in farming both man and nature conspire hourly to thwart
the agrarian. The farmer's plight is a physical contest in which he
can see, smell, hear, and thus be one with his tormentors any hour
he pleases. To the Greeks—who believed in the acquisition of
knowledge through pain—Aeschylus' *pathei mathos*—farming ac-
cordingly became, in Xenophon's words, the "best tester of good
and bad men," the clearest mechanism known to teach the man of
the *polis* the properly tragic view of the universe.

When your raisins are ruined in an hour, your pears rotted by
coddling moths, you indeed understand that we do fail. There are
no second chances and the race does not always go to the swift. So
Virgil glumly wrote of the farmer's travail that "it is a law of nature
that makes all things go to the bad, to lose ground, and to fall
away." To understand us, you must learn, as we have, the potency
of our enemies. As Mr. Crèvecoeur put it:

> Now if you unite the damages which we yearly suffer from all
> these enemies, to the badness of our fences, to the want of

subordinate workmen, to the high price of our labour, to the
ignorance of our tradesmen, to the severities of our winters,
to the great labours we must undergo, to the celerity with
which the rapid seasons hurry all our rural operations,
you'll have a more complete idea of our situation as farmers
than you had before.

If farmers possess humility, endurance, and perspective, it is
because they must if they are to survive their wars. Each adversary
tries the farmer in a unique way. The need to accept the tragic limitations of man derives from the inexplicable damage of the
viruses; insects teach us that war, strife, and aggression are with us
always—surrender or pacifism gets us and all we hold dear killed.
And because we square off against mites and mildew each year in
a war that has no beginning and no end, we understand that
progress, success even, is transitory and quickly lost should we let
down our guard. Farmers, if they have farmed for any length of
time, also gain historical perspective in which the present is seen
as not particularly any more important than the past. When I lost
my first raisin crop, when my first harvest of table grapes rotted, in
my hysteria I sought to instill similar panic in my parents (who, after all, had cosigned the crop loan). But they scoffed, and were
serene not frenzied, rattling off similar disasters of the last half
century—the '32 frost, the sibling who during the war demanded
to be bought out with cash, the '47 price collapse, the big rain of
1958—finishing with not much more of a reply than "Well, that's
farming; next year will be better." I had not known their wars, but
at that moment I saw how their campaigning had made them so
contrary to their peers in town. How strange that the grubby
farmer in the coffee shop, the sullen agrarian in line at the post office, is in fact a decorated veteran whose survival is proof that he

has won more than he has lost, that at year's end he did harvest his crops after all, that he is a warrior on the front line whose combat has made him wise and different from the rest of us.

I. Viruses and Bacteria

The farmer's worst antagonists among the lower animal kingdom, as with the adversaries of society at large, are and have always been invisible to the naked eye. They have no color, no shape, no beauty. They emit no sound. Few make any movement we can see. You see only their damage, not them. Most viral, bacterial, fungal, or parasitic intruders ultimately kill their hosts: our trees and vines. But they are also programmed, thank God, so that their success is also their eventual demise. They at least leave you when your orchard is dead. Farmers do not really know the origins of these pathogens that they cannot detect, much less their life cycles and methods of reproduction. No wonder the Aristotelian Theophrastus, some twenty-four centuries prior to us, despaired of the work of the invisible tree and vine killers. He dubbed them the bringers of the "unnatural and violent" way of death. True, we no longer, as the Greeks did, call their invisible work "unnatural" and "violent." But do we know much more than Theophrastus about their causation and etiology?

I can at least recognize the symptomatology of the sudden and hated microscopic attack on agriculture: burned pear limbs, the dropping of grape leaves, the malformation of vine canes, or the ugly saplike mucus that oozes from the now cracked scaffolds of the plum tree. Plant pathologists, who have identified such infectious villains with the aid of the electron microscope and the genetic marker, and thus have given them a nomenclature of genus and species—*Agrobacterium tumefaciens, Phomopsis viticola,*

Fomes ignarus, Xiphinema index, and the like—know little of how to stop them. They offer only concerned advice on how to avoid their alighting, how to curb their outbreak, but never can say how to end their destruction once they have arrived. The microbes and submicrobes are the smallest of the farmer's enemies and thus logically his most awful. The educated in the university and research stations assure us that these eaters are alive. But they might as well be the stuff of superstition, unseen humors, vapors, or witches' curses for all we can see of them. The Greek poet Hesiod put their infection among the worst of evils that flew from Pandora's jar—those "diseases that come among men in day, and by night, and come upon them on their own accord, bearing evils to men in silence."

Just as modern medicine has largely failed to stop the baneful agents of silence such as cancer, multiple sclerosis, and Lou Gehrig's disease, so too Pierce's disease, Spanish measles, phylloxera, bacterial gummosis, black line, and fire blight that ruin orchards and vineyards remain unconquerable. When a farmer sees his tree blacken, or his green vine suddenly scorched, he can neither explain nor react to the evil he cannot see, feel, hear, or smell. Perhaps for the first time in his life he understands why one of his daughter's legs suddenly goes limp with paralysis. He now has a hunch why years later another develops a metastasizing tumor in a healthy breast. He sees at last why mute plant and human are, in fact, inseparable, and thus subject to the same random but cruel and inescapable doom of the universe.

So we humans, like the tree and vine, become maimed and die, for no apparent reason, from no perceptible cause. As we suffer firsthand from the enemies of agriculture, we of the countryside do not believe the current dogma that the universe is a happy and kind place rather than a brutal and tragic ephemeral abode, where there are still limits to reason and the power of rational explication. We do not believe that the march of science can explain everything; our farm shows us why reason cannot be God. We expect, rather than are surprised at, the sudden arrival of the wasting illnesses that we cannot combat nor describe nor explain. We, who do try to live by reason, nevertheless have learned to make room for the inexplicable. Why not a daughter of twenty gone, when whole verdant vineyards wither? If plums are to split, peaches to rot from within, the insides of beautiful pears to be devoured by the unseen fungus, why not a son crushed under the wheel, his arteries sliced by the gang knife, or his blood full of odd-shaped toxic cells? We out here shrug when our walnut orchards blacken and die, half-expecting our apricot orchards to rot every third or fourth year on their own respective and private schedules. We who depend on nature are utterly ignorant of its logic—if indeed it has even a logic, much less a fairness.

So we dour farmers have little faith that therapy and counseling can ameliorate man's life on earth, which is largely, like his peaches and plums, to be infected, maimed, and aged by nature before at last falling away for good, dropping to the ground in his own proper season. We have little assurance that the philosophers ensconced in our universities can ever really tell us who we are or why we are here, much less where we are going, and for what reason we are to go on. We look instead to our trees and vines and so learn only that we are in a life-and-death struggle with nature, which eventually always has its way and goes on when we finally cannot. A morally neutral monster that neither punishes nor rewards but only exists oblivious to and larger than us; to our petty minds it is so cruelly indifferent, so immune even to our science and philosophy. Given the rigged rules of that game, we farmers also do not believe that technology's final conquest of the microbe is just on the horizon—or that even if it is in our future, it is necessarily an entirely good thing. No, we learn only how to accept, not change, fate; to understand, not to reinvent, man; to seek out faith, when reason is exhausted. I learned more about how to understand Greek tragedy from the neighbor across the road who lost his livelihood in a single day than from all the graduate seminars in Greek drama put together.

In general, as again in the case of man, it is only the visible and concrete foe in farming that can be fought and sometimes checked. Perceptible, manifest mayhem by man and nature—a gusty wind, a swerving tractor, an untrained pruner, a voracious family of rabbits—is usually not so fatal to the tree or vine. Such damages—riotous and arrogant though their assaults may be—remain tiny affronts to the farmer's march to the harvest. They are on the cosmic scale something minor and therefore curable. They rarely injure our livelihood. Just as medical man can reconstruct a charred hand, reattach a severed arm, remove a fetid appendix, or transplant a liver, so too in farming we chainsaw off dying branches. We restore parched vines, tar chewed bark, and prop up sagging limbs.

But the tiny and invisible enemies of agriculture? Once they infect, they kill. They kill until they are spent and can destroy no more.

I have bragged on my skill in poisoning gophers. I have shot scores of rodents and blackbirds in the last thirty-five years. I put tree wax on broken plum limbs, and feel smug when I chase off legions of lost and innocent suburbanite horsemen. But I have said not a word to anyone about the hundreds of trees and vines that have gone to the dozer or chain saw due to infectious things I could neither see nor understand.

In what we might call the first category of the farmer's enemies belongs the bacterial fire blight of the orchard, which is at home equally among apples, pears, and quince. This bacterium burrows

in the bloom during an especially wet April. Within a few months it
burns entire limbs off. As it descends into the stump, it will kill the
life force of the tree itself. In this great age of antibiotics, farmers are to think that all bacteria are conquerable. They are advised during bloom to spray on
the orchard industrial-strength Terramycin, tetracycline, strepto-
mycin, and other conglomerations and mixtures now dubbed
with fancy trade names like Agromycin. In fact, farmers spray
such medicinal mists four or five times during the ten-day bloom
period. They drench the grove when the pear orchard's defenses
are down and its immune system exposed to the elements through
its beautiful but naked flowers. But fire blight is not like the child's
earaches or tonsillitis, but more like the unconquerable flesh-
eating streptococcus bacteria of the tabloids that bite deep to the
bone and quickly blacken sinew and tendon. When fire blight gets
in and infects a tree, it slaughters. Farmers call in saws one year,
the dozer the next. No wonder this family has always had an affinity for surgeons, the sawbones who are decisive and know what
and why they are cutting, so different from the oncologists who
use strange brews in vain to poison and kill the cancer a day ahead
of the host. My poor grandmother was mystified—and then re-
pulsed—by the onset of chemotherapy; far better, she declared, to
"cut out" a cancer and thereby give the stricken an extra year or so
by lopping a nose or limb off.

   The dutiful and neurotic farmer cannot be satisfied with just
the prophylactic of antibiotic-laced spray at bloom. On the theory
that the bacterium does not appear *ex nihilo* (to wit, the Greek
atomists' "nothing can be created out of nothing"), the farmer also
hunts down his unseen enemy throughout the year with a variety
of weapons in his mechanical and chemical panoply. The confi-
dent, all-knowing aboriculturist empties his arsenal in a shotgun
approach at the invisible pathogen.

And, don't forget, we farmers do have quite an armory, we late-twentieth-century poisoners of the wild. Now in November the farmer sprays copper on the trees. His theory is that the change of season, the dropping of leaves, and the transition toward dormancy might leave the tree vulnerable to infection—that a blanket cloud of metallic antiseptic might smother any unseen airborne bacterium lurking in the immediate atmosphere. Now in January his pruners dip their pruning shears in disinfectant bleach—worried that without prophylactics their own metal cutters in the dormancy of winter might still transmit the unseen toxin secretly from a diseased limb onto a healthy tree.

Is not the farmer right that the bacterium is always present? Yes. So, again, in February the yeoman burns any infected wood he spots, from small twig to the stump itself. His assumption is that infected prunings can radiate the bacterium onto healthy stock in its midst. We must quarantine tree from tree. In the end, if bacteria are there—and they are always there—if they feel at home in your particular environment, your weather, your cultural practices, your species of apple or pear, they will surely kill your trees in the spring no matter what the preventative at your disposal.

"We should have planted more quince," I boasted to my brothers in 1988 when for the fourth year in a row, while raisins were no good, while plums worse, while table grapes remained unsold in the broker's cold storage, we had a bumper crop of quince at good prices on our tiny one-acre grove. "You should never have planted quince in this Valley," most scoffed in 1989 when fire blight without warning hit the orchard and we sawed off and burned the top limbs of the trees. The pragmatic and self-educated grower recognizes the lethality of this species of killer far better than the well-read and smug bacteriologist. The former judges by what he sees, not, as does the latter, by what he reads or is told. Ernie DeLeon taught me that all my research journals and tree-fruit books were of no value, that my education was but a nodding plume, when he studied the annual charred limbs of my pear orchard and offered unsolicited advice: "Get those trees out now; they'll just burn up each year."

And they did burn. Columella two millennia ago said farming knowledge does not come from books but from praxis. So we farmers listen to the Ernie DeLeons of this world, whose heads are clear of books, but whose eyes are on leaf and limb alone. Degrees, the bacterial foes of agriculture taught me, do not ensure knowledge. The pear cares little whether the concoction that stopped the blight was dreamed up by a Ph.D. or an immigrant from Oaxaca.

But how lethal really are these microscopic foes, the most malignant of the natural adversaries of agriculture that the farmer assumes are always there to destroy him invisibly? Do these viruses and bacteria kill all the farmer's vines, annihilate his vineyard, destroy his orchard, put him out of business? They do not. Are they what was responsible for the agrarian cataclysm of the small American farmer in the 1980s and 1990s, this age-old bacterial and viral onslaught that has neither prophylactic nor cure? They are not. At least in this Valley they are as yet not the potato famines of old that put an entire people in the grave or on the boat. Our government need not have a War on Blight to save the small farmer. Ninety percent of California's farmland now in the hands of 5 percent of the landowners was not due to the triumph of *bacterium terribile.*

Pierce's disease, gummosis, fire blight, and their brood—the most lethal of nature's tree and vine pests—again like cancer and MS, do not often annihilate families or cease human progress. They simply turn the life of the individual bad. They make living itself nonsensical, inexplicable, and ultimately tragic. These pathogens kill when they should not, and so destroy the psyche, not the entire harvest. They often murder not the weak and diseased flora, the unproductive, and thus the marginal tree—horrible enough, yet with a gruesome logic to it. Rather, like their foul counterparts that infect man, they slaughter the strong and the healthy vine, the fertile, young, and the irreplaceable tree—and not thousands either, but enough nonetheless. What else is it but a tragedy to see the best of a Santa Rosa plum orchard ruined in its prime by bacterial gummosis, limbs scorched, tiny plums mummified, the entire enterprise of cultivated trees itself a ruin?

In 1979 our small, beautiful plum orchard was charred in the space of a few weeks as the contagion took hold. A beautiful white blossom in the February orchard, deep green young leaves in the March breeze, a good April plum set—and then death by June as leaves and undeveloped plums littered the ground. Unproductive tree varieties nearby remained unscathed.

"You're spreading the gummosis down the orchard row with the disk and the tractor," the farm advisor warned. The beautiful Santa Rosa trees died. "Kill those gophers and squirrels, they're the ones getting the canker into the roots," the university pundit lectured. More trees died. "It's the nematodes in your sandy soil that wound the roots and make the gummosis worse," the neighbor concluded. Still more trees died. "Uproot the dead trees, fumigate the ground, spray copper each fall and spring, prune in the summer after harvest, never in the winter," we read in the pathology handbooks. All the trees left died. "Your grandfather had the same problem; his whole orchard died in the thirties; the ground is just too sandy, no good for plums, too good for gummosis. You should never have planted plums there," my late uncle offered as ad hoc postmortem. The dying trunks all turned a final black.

The counterassaults against the gummy bacterial canker were not unlike the concurrent family battle against the brain tumor of my mother. We were then, we also thought, prepared. "A robust chemotherapy can stop this reoccurring meningioma," the smug oncologist advised. "Are you willing to go all the way with an aggressive radiation program?" the young radiologist demanded. "I can take this tumor out two or three times more if I have to," the veteran and dexterous neurosurgeon assured. "We have an entirely new curative protocol at Berkeley," the university cutting-edge specialist bragged. "I still would prefer to practice aggressive medicine on your mother," the now not-so-cocky neurologist meekly offered as the growth devoured brain and spinal tissue.

In the end, like the gummosis and our own reactive spraying, cutting, and burning, the brain tumor, we knew, always reappeared and grew bigger. The more the medics drugged, cut, and radiated, the bigger, more foul, and ultimately more lethal the tumor in my mother's brain became. The doctors, I came to learn, were not unlike the chemical salesmen who sold us the new and most improved brand of antibiotic: suited and tied, a vocabulary full of babble, with a polished veneer to mask the absurdity of their chemical soups and pharmaceutical concoctions and the utter feebleness of their regimen. I had respect only for the bloody surgeons, the cutters, who like us pruners at least went in and took out what evil they could see. Ultimately, because I had seen gummosis do its work, I told the oncologist, "No more, you're through. Show's over. She's going home." Exhausted like our Santa Rosa trees, she too died eleven days later on the farm where she was born, about two hundred yards to the east of the dying plum orchard itself.

Gummosis in that plum orchard did not kill us or break us that year, as it had not done so in other apricot and plum groves on our farm—as one hundred twenty years ago it had not killed or broken my mother's ancestors who before us worked on this same infected soil, the same forebears who I guess likewise had passed to her—and every female in that line—the gene for the breast tumor and its accompanying brain cancer. But that tiny pathogen did make life hard. Gummosis taught us how ugly it is to look at the dying, made it futile to think of the lost beauty, of what might have been gained in years without the inexplicable infection; those withered limbs taught us to accept what was unfair, without reason, and ultimately unstoppable.

The invisible enemies of agriculture do drive the farmer a little closer to the edge. They do convince him that there are unseen vapors and dark humors in the universe that are soon to give him misery even as he sleeps, which are really not explicable or controllable through the antiseptic books and protocols of the medical class. Viruses and bacteria do prove that ultimately we, like trees and vines, are here to suffer on this earth when random disease comes when and where it should not. In agriculture and in life, some of the worst of man's animal enemies, I know now, are those that he cannot fathom and so cannot fight. We farmers know, as you should also understand, that the seemingly invisible is there and it will come and it will take the best of us when it should not. And so we wait, weak and hopeless.

Yes, we wait. But we prepare, as others should, and we grimly resign ourselves to that rotten limb and blackened trunk nonetheless.

II. Fungi

Bacteria and viruses are not the only invisible diseases that prey on the harvests of trees and vines. The others are the fungi, such as powdery mildew and botrytis, that are still not so discernible to the eye. But may I be brief in their description, for they are foul, yet petty, things that ultimately but ooze and drip? Despite all their rank and fetid accomplishment, fruit fungi do not even kill the host, but rather rot and discolor the produce alone. Noisome and ugly is the work of mildew and rot. They are sloppy and they stink. The harvest of "the vinegar vine" is what the Greeks said such grapes become.

Fungi operate in tandem with insects and precipitation: cracking the skin and swelling the insides of the soft-skinned fruit, where lesions are to be expanded and cultivated by gnats and rain, leaving vegetative pus and foul slime in their wake. "I don't have to see your goddamn rot," the field man for the shipper said of our rotting red grapes, "I can smell the sonofabitches half a mile away." He really could. And while he was a foul thing himself, he was on this rare occasion not a liar.

These fungi are like the poor graffiti artists, pimps, whores, the homeless in the street, the panhandlers and runaways who take most visibly but not most effectively from civilization, whose excrement, vomit, and offal are on the curb and sidewalk, and in the park, but who do not bomb or machine-gun, nor with suit and ties plunder the nation's wealth over computers. Yes, they are ugly and they do reek. But alone they cannot reach the vital organs of their host. They are after grapes, not the vines themselves. They may piss and puke on the curb outside the lawyer's office or city hall, but their power to do evil, I confess, pales in comparison with those within.

Unlike gummosis or fire blight, the farmer, if he should get angry enough, can control the fungi on his fruit through natural toxins themselves. Elements deep within the earth, cheap and nontoxic, like sulfur and lime, desiccate the cracked fruit surface, suck up and neutralize the detritus even as they extinguish the budding spores of infection. With your whirling duster perched behind the tractor, any of you can coat thousands of vines with clays and ores in an afternoon. Goggled and scarved, like some strange bug himself, the farmer on his mechanized duster emits vast clouds of sulfur, copper, or lime.

These primeval dusts settle on millions of berries within his vineyard. One encompassing prophylactic shield of fire-and-brimstone dust kills spores of every sort, particulates so condensed that they change the very dynamics of the vineyard atmosphere itself. No, if your August grape field stinks and runs, it is mostly your fault for not being out on the godforsaken duster the prior April. If you have but the will and patience, the fungi can be periodically swept away. Since that sulfur mist burns your eyes, causes fire even in the mucus of your nose, you rest assured that it will burn up spores who are less than you.

Pathogens of soft summer fruit other than mildew—brown rot, bunch rot, and leaf curl—that get through the first line of prophylactic dust, beget stink and slime. Again, they are unsightly. But even at their very worst you lose the crop, not your tree and vine, to fungi. A year, not a decade, is gone. They smell up an entire row of vines. But they do not turn the leaves brown, the stump black. The stinky and oozy, as is the rule off the farm as well, assault your eyes and nose, less so your blood and flesh. Infected fruit turns away the urban shopper, but even if consumed, it rarely turns the stomach.

I pass on the scientist's improved answer to such rot, the next-generation antidote to replace the primeval dust that is mined out of the earth. Let it be said only that space-age chemical fungicides that supplant natural dusts have a predictable cycle: they work wonderfully at the outset as the targeted mildew itself becomes an endangered species; they shortly incur natural resistance as mildew in its new and improved strain creeps back into the vineyard; they are next rendered useless as a new superrace of mildew rages as never before and must finally be superseded by entirely new and more expensive fungicides to the greater profit of those who ostensibly are there to help the agrarian.

Thus the farmer parks his old sulfur machine or copper duster in his shed, and buys the latest hyped and advertised high-priced chemical. Five years later, beset with abject chemical failure, his body full of carcinogens, he drags his clanky contraption back out of the barn. Now he replaces its archaic bearings and rotten rubber belts, and begins with his ancient machine anew to apply safe and primeval dusts to his new fungicide-resistant and rotting grapes—a process that has no beginning or end. The moral? To kill fungi farmers simply use the age-old ores of this earth. The battle against the foul is not one to tinker with and contrive, but simply to have the will to use what is tried of the ages and found unfailing.

As long as a single grape farmer exists, he will despise blasting on that infernal sulfur dust that scorches his eyes and yet causes no cancer. And American farmers will also welcome each new generation of wettable chemical powder that blows cleanly onto their vines, its potency against fungi weakening even as its cancer-causing properties among mammals becomes clearer.

Rot on ripe fruit is odorous and embarrassing to the vineyard. But its lasting effects are not lethal, and both man and nature provide ample antidotes to its spores. Citizens, if we have the patience to continue with the dirty, tiring, and unending counterattack of this world, we can stop the ooze. The rub is the sheer boredom and disappointment in accepting that the solution is not new, but a matter of patience or rote, not brilliance or innovation. A year ago four thugs pulled off the road and began robbing our oranges in front of the house. They were tattooed, petty gangsters, tossing oranges into plastic garbage bags, while blowing weed and throwing bottles on the pavement to the worst music imaginable. I eyed them from a hundred yards away, clinging to the old myths that perhaps they are just frisky high-school students, and do we really need all those oranges?, and are they any different from brokers?, and maybe they will come to the door and pay, after all? That stream-of-consciousness lie lasted about one second. And then I realized, no, they are thieves, and they will come back tonight, tomorrow, all season, unless confronted and stopped. And so I walked out to them, thinking—as I used to on the way to the sulfur machine—"Here it goes again...."

III. Flyers, Hoppers, and Crawlers

The pests of biblical magnitude are, of course, arthropods, winged and legged, on rare occasion seen and also scarcely discernible. The phylum is the most fascinating of agriculture's enemies. Its members can with difficulty be observed as they prowl—and sometimes even heard as they buzz, gorge, suck, and spit in their millions. These worms, spiders, wasps, mites, moths, and hoppers arrive in your field in a variety of bizarre shapes, sizes, and colors: black, brown, orange, and yellow, pinhead in girth to the size of your thumb, from innocuous to the touch to venomous and near lethal in their bite. Their strength lies in their sheer multitude, inasmuch as it is easy to kill a few thousand while millions are oblivious. Their weakness is that they are alive in the rawest sense. Thus, as visible members of the animal kingdom, they do need, as we do, air, food, and water. The bug clan has both pore and mouth—and so it can be killed if poisons are deposited in either of those apertures.

Luckily for us, since antiquity there have always been known toxins—peppers, juices, lime, and dusts—for all of their orifices. The trick is merely one of logistics, not of poison per se: providing the dose in enough quantity and strength to kill the swarm before it demolishes the vineyard in question. The problem for Greek and Roman farmers was not the absence of poison, but the lack of tractors and sprayers that blast hundreds of gallons of the toxin each hour at the proper hour onto the vines. The ancients could kill thousands of insects and spiders, even as millions took their crops. One Power Takeoff–driven pump, agitator, and fan would have been worth a thousand Thracian slaves toiling in the Po Valley.

Before the nightmarish discoveries of the Nazis—that is, during the nineteenth-century infancy of American viticulture and arboriculture—the farmer's chemical counterattack was more elemental: essentially a mere update of tactics culled from Greek and Roman treatises on insect eradication. The general rule was clear: anything known to sicken or kill man himself was used in lower doses to kill insects. These mammalian toxins were logically far more lethal to those smaller central nervous systems, whose blood was not red, but white, brown, and black. Nicotine, lead, cyanide, and arsenic—the unholy quartet that our government now spends billions to wipe out from our walls, waters, and lungs—do kill bugs of any kind. Always have, always will. Of each of those poisons Uncle Tilford said, if it kills your lungs, it will kill hoppers; if it kills those on death row, it will kill mites; if it kills syphilis, it will surely kill bunch rot.

Before the age of the tractor-driven sprayer, to stop vine-hoppers from denuding his vineyard, my grandfather, together with his brother, father, and uncles, would strap small metal tanks on their backs, replete with hand pumpers and spray wands, and walk the vine rows. From photos, diaries, and oral history I gather that day in and day out, twelve hours and more each cycle, their tiny bellows would sprinkle nicotine on the leaves. Old Uncle Tilford used to visit to confirm some of the story ad nauseam.

Your grandfather, my goddamn brother-in-law [who has been dead now twenty years], and I blew on the nicotine all spring of 1913. But he got sick and couldn't stand the smell. Hell, I've been smoking since I was nine, and the job was not much more than a pack or two of cigarettes for me.

When Tilford proceeded to the subsequent chapter on arsenic and cyanide and his own singular propensity for vineyard heroics of eight decades past, I politely but ambiguously used to ask five-foot-one-inch, chain-smoking, foul-mouthed, Stetson-hatted, alligator-booted, nine-decade-old-and-more Uncle Tilford, "Why aren't you dead yet?"

Cyanide, arsenic, and frightening oils of lead derivation lay in hibernation in the barn, long forgotten and unused when I came of age. What to do with the venerable breed of poisons that are not of complex and artificial creation, but agelessly toxic in their elemental form? Bury, burn, or dump them? Confess to the public EPA officer that your vineyard outside the front door, the soil itself that rests above your water well, for eighty years was the daily repository of the debris from such an alkaline war? Do you curse your predecessors for putting powdered death onto your inherited earth?

You should, but more often do not. You read in their diaries that the poison drenched them as much as it did the insects, given the primitive state of mass application and the once macho ethos of spraying bare-sleeved without protection. You heard that the purchase of children's shoes and food depended on the ability of the trusty cyanide to knock down mites. Costly nicotine stopped the swarms of hoppers and so also led to clothes and food. What was left in the bottom of the can went under the house to get the termites, or was thrown in the stagnant pools to stop the disease-carrying mosquitoes. No, more often, I confess I am rather appreciative of the remnants of these bags and cans of ancient powders that once kept the farm alive at the century's beginning, that killed bugs perhaps more often than they destroyed my ancestors. They are awful toxins. Yet they command a certain respect, as artifacts of salvation every bit as venerable as the 120-year-old wagon on the hill or the rusted mule-driven scraper in the shed.

But I said earlier "before the Nazis," in bitter irony and in poor taste, of course. Their nefarious mastery of genocidal gases was soon to be applicable to the benefit of man through the extinction of insects in the vineyard and orchard. From their loathsome notebooks arose in the 1930s the organochlorides and organophosphates like DDT and parathion, when the work of such monsters was adapted by benign nerds in American laboratories, who sought to gas billions of insects rather than millions of innocent people. Those first-generation nervous-system caustics, of course, were deadly. They sometimes killed any who breathed their mists, had their arms soaked, or eyes blasted. A neighbor a few miles away in the 1950s was found crawling on all fours, doglike in delirium, after a day on the parathion rig. He was in and out of the insane asylum until he died. Organophosphates killed every living thing in the orchard and vineyard: mites, hoppers, and worms, but also predator spiders, wasps, and ladybugs, even birds, gophers, and snakes—and the occasional laborer. Even the lowly toad snug in his hole beneath the vine was not safe. I have seen these cute amphibians stiff and yellow outside their holes on the day after the spray. Agriculture's caustics—made infamous by Rachel Carson's inflammatory rhetoric—created the whole science of "integrated pest management" and "organic" farming, the desire to "discourage" pests without chemicals, to use natural irritants, cultivation practices, friendly insects, and muscular labor to grow plants in peace among bugs without recourse to poison.

Why, you ask, would any sane man, as I have, apply even the less toxic ancestors of parathion—Lannate, Guthion, dimethoate, lindane, carbaryl, Thiodan, or diazinon—to kill everything on leaf and tendril? Could not, instead, natural predators be released without your lungs getting tight or your arms and throat getting scratchy? Could not organic juices from garlic and pepper be substituted for chemicals? Oils and soaps found to smother a few rather than kill all pests? Insect habitats altered through hoe and shovel? Might even fruit be sold, slightly chewed, a bit sticky? Why pesticides? Why poisons of the central nervous system? Why?

They kill, kill, kill, and kill some more.

Take a look at a mite under the magnifying glass, or a roller worm or coddling moth. They are not pleasant-looking creatures, but spiders whose entire design is built around their mouths. No wonder that, blown-up, they make adequate science-fiction monsters on the screen. These are enemies whose ferocity is matched by their grimness, and they require poison whose lethal stink matches their own.

If you dust the organic bacteria Dipel on grape leaves, it will deter, but not stop, a grape-leaf skeletonizer or coddling moth. If, as one should, you sow cover crop, plant berries and flowers, and use manure, you can grow predators to chomp on, but not eradicate, deleterious insects. Is it to be hours of contemplation, repeated application, and constant monitoring, or one shot of caustic Omite that will kill red spider mite for two months? The choice is not really the blinkered farmer's alone. The decision has already been made by the environmentally correct consumer: he wants fruit plentiful, colorful, hard, and fresh, free from crack and scar—not high-priced and tasty, overripe, leaking, and pock-marked, a hitchhiking gnat or stowaway fly now swirling amid the glistening produce section at the local supermarket. Under no condition must the vine-hopper's nontoxic excrement dot the grape. An English professor once called me complaining about the grapes he bought at the local organic farmer's market on campus. "They have black dots all over the bunches," he raved.

"That's hopper crap," I replied.

"Well, who wants to eat it? Not me," he fumed.

The farmer usually employs the strongest, cheapest poison that lasts the longest without killing him outright or giving his children too likely a chance of getting cancer within the next decade. How else can you get grapes, green, hard, smooth, pretty, big, on time, and out of season to New York from Selma? When the suburbanite gives up his chemically green lawn—requiring toxins often outlawed in agriculture—and can live with crabgrass, dandelions, and an occasional sand burr, then we farmers will have a consumer market that demands organic, ugly, perishable—and delicious—fruit. The moment a consumer wishes an out-of-state, out-of-season peach or grape, he has voted, inadvertently or not, for agribusiness's arsenal of chemical preservatives and prophylactics.

I once sprayed dimethoate, and in between the stinking loads, as the five-hundred-gallon spray tank refilled, I read the *Inferno*, wondering whether I was in Hell or earning my way. I have dusted organic Dipel six times on vines and watched worms sicken but not die. I have sprayed the toxin Lannate once and watched them drop off before the tractor left the field. Yes, we know it is legal but wrong. Yes, we know it is expedient but nonsustainable for the millennia. Yes, we know that it costs money we don't have. Yes, we know there will be nemeses to confront for doing what we should not; we know that he who pollutes his land must atone tenfold in the hereafter. But farmers as a last resort use terrible chemicals because they do kill, no questions asked. And there are some times in farmers' lives when there are insects, millions of them, that must be killed and killed quickly if their brethren are to eat cheaply and plentifully and on schedule—if one man is to feed ninety-nine other Americans 3,000 miles away by next Tuesday.

The consumer too can stop the use of poison when he pays more—and pays that more for taste, not appearance. Until then we farmers continue to poison our fruit, ourselves, perhaps you too. We will because it is legal, though not moral, because we are told it is absolutely safe and tested, which is untrue, because it brings us short-term salvation, when it should not, because you trust us when you must not—and because you want fruit of a type and at a time you should not.

Mites and hoppers breed geometrically, not arithmetically. Their reproductive cycles overlap, as populations go from hundreds to thousands to millions to billions in a vineyard in only days. A few grape leaves are brown from mite in the morning. The next evening the grape row itself is now no longer green. By noon the third day the entire vineyard is burned and an ugly brown. Even the most diehard organicist knows that when his carefully constructed equilibrium goes out of kilter, when his natural regimen is askew, there are always tons of poisons but a few miles away in town that can relieve him of his disaster in hours, and so save him from the bank. Reader, please remember that vineyards and orchards are not nature's creation but man's, these big insect dinners that are tasty and so unnaturally all gathered in one place for the taking.

The farmer also learns of this world from the billions of arthropods, just as he has learned from viruses and fungi and the other less visible foes that are his teacher. But the lasting lesson from war against the bug and spider is somewhat different. They and their progeny are insolently there in your vineyard and orchard. Their ancestors—in a direct line no less—were here when yours were too. Your ninety acres of vineyard is a family heirloom of the mite and hopper clan, where a thousand generations and more have moved not more than a few feet from your vines. The mite clan owns your vines as much as you, always has and always will. Their descendants do not move from their ancestral abode.

Insects and spiders land on you, sting and bite you, and fly and hop on your work. Before your very eyes they eat what you grow, always, forever, while you are awake, during your sleep. They will crawl, fly, defecate, and consume hourly until they are stopped, and they can at the millennium, if you but have the will, be stopped. Otherwise, they will devour what you have raised and hold dear in the process, what the bankers rightly say is already theirs for the money you have borrowed. Unlike the insidious and stealthy "unnatural" viruses, you see the excrement of these tiny animals, notice their damage as they munch undisturbed, and so you realize again that your work is their fuel, their lifeblood.

They must eat what feeds you or die themselves. Their lesson for us all? It is not the more metaphysical and profound one of inexplicable fate and tragedy learned from the invisible and not understood viruses and bacteria that kills from within. Nor is their messy sermon like that of the fungi—that the unwanted are merely bothersome and odorous but not lethal. The fight to kill the spider, worm, and insect is more elemental and real, as are the ugly bugs themselves. Tattered leaves and plum holes, sticky dung on the grape bunches—an entire load of fruit rejected for "blemishes"—not a mysterious wasting debilitation of the plant, are their trail.

The arthropods demonstrate visibly to you that there are millions of brainless enemies in the world that want what you have—and have the mouths and numbers to take it. Their hunger is not born out of evil. It is existential and a part of their genetic code itself. Their destruction has no real logic—after all, insects will finally eat their own food supply into oblivion. Their lives have, of course, no morality—bugs do not choose to hatch on and eat your plum leaves instead of the neighbor's because of a fib on your income tax or a shameless abandonment of your wife and children. They are morally neutral, like Nature herself. Their existence has no beginning and no end—a thousand years from now insects will appear from nowhere in an orchard, like Aristotle's spontaneously generating beasts themselves. Until recently, I never knew what my grandfather meant in 1963 when he said, "Well, we'll see if you boys can hold this ranch, when others come to take it."

"Others?" "To take it?" How absurd! Only later when I entered my own war against the arthropods did I agree that these winged and crawling enemies are but smaller versions of man himself, who predictably and naturally struggle to grab what others have. People in Fresno have ridiculed me for spraying bugs, for getting sick from accidental doses of dimethoate, Dibrom, and Omite, but they have never asked why I took on such an unpleasant job in the first place. Or how they thought it possible for New Yorkers to eat fresh apricots in early May.

From the lowly bug the farmer understands that forces in the world indeed do exist to breed, eat, and ruin his culture until stopped. He extrapolates—often wrongly, you more educated no doubt object—from the insect, spider, and worm and transfers that mundane experience with these small adversaries of agriculture to the larger and abstract world. I think the reductionist farmer alone, almost ridiculously so, understands the mind of his nation's enemy far better than the professor or diplomat. No, the farmer as fighter says quite honestly and without censure to drop the bomb on Hiroshima if that is what it takes to stop the onslaught of those who hate us and would shoot us down and would behead and torture millions of the weaker.

Of course, the farmer is simplistic. Of course, the man who grows plants turns out to be not entirely a nurturer, but rather more often a destroyer as well. Of course, in his instinct for self-preservation, the farmer is himself dangerous and to be watched. On examination, he is a frontline soldier who understands the attacker and the force needed to kill an enemy, who will unthinkingly eat you and the innocent weaker until smothered, burned, or poisoned. This boorish farmer, slave to his so deductive, infantile logic, would advise the Harvard-educated diplomat of the Rwanda holocaust: "Get some troops over there quick to shoot a thousand or so of those murdering bastards and then the other million innocent still might make it." I think the farmer would offer the diplomat the absurd advice that there were only a set number of real killers in Rwanda, evil thugs who must be slain immediately before their cowardly and vicious rampage convinces the weaker bullies to join in the blood lust. And the farmer would say to the university professor, forget for a moment about trying repulsive Mr. Pinochet, the killer of a few thousand, and instead first extradite the architects of the Cambodian holocaust, those in comfortable retirement, who murdered over a million with impunity.

I have put down spider mite easily the very day it flared up, with a variety of toxins; the one time I waited (thinking "I bet it may just quit on its own"), the vineyard was burned up in three days—on day four I sprayed to kill thousands of tiny spiders as millions kept sucking unnoticed. Perhaps with the advance of "peace resolution theory" in the century to come, vine mites may yet be persuaded to file down their fangs; perhaps mildew shall be coaxed to cease its rotten ways; perhaps both will promise to be nice fellows and so take just a quarter of the crop.

The farmer has a Hobbesian view that, as on the land, so elsewhere there must be constant attack and retreat, victory and defeat in the effort to survive one more day. Like Plato's Cretan stranger, the farmer believes peace is but a parenthesis; war is the natural state of things. Like Aristotle, he knows that the stronger creature forever tries to acquire from the weaker in a war of all against all. The farmer is not of the Enlightenment mind, which suggests that with more logic and reason the enemy might yet see the light. He is no Kant. Nor on the other extreme is he the dummy Romantic, who, like Rousseau, thought that had we but less convention and culture, we might yet revert to our natural harmony. And he is not even the Judeo-Christian who tells us evil in the world is merely the work of the Devil and his gang of nonbelievers, who can be prayed and confessed away.

Rather, the farmer is the true Greek, Heracleitus in the flesh, who says that war is the father, the king of us all, Thucydides' "harsh schoolmaster" that reveals us for what we are. The diluted-spray, the delayed-spray, the no-spray brings on the insect—the bug that will eat until it can eat no more. The world is not a choice between savagery and utopia, but rather between savagery and far less savagery—so says this reluctant farmer who sprays yet would rather not be on his costly leaking, stinking, and toxic rig at three in the morning.

But what of altruism, you say, what of the voice in the wilderness that says all such violence is not as it should be?

To show mercy, as we often must, the farmer retorts, first recognize and understand evil, lest a mercy that soothes the conscience and makes us smug creates greater misery for us and others than before. On the farm the divide between leniency and laxity is tenuous and prone to be misunderstood by both man and animal. The farmer, like old Edmund Burke, realizes that forbearance without a limit is no virtue. Clemency, the bug-fighter has now learned, can just as often kill the weak should the evil and the stronger be let loose. When at war, when the moral capital tips your way, when the enemy does wish you dead—or simply to stand aside while it moves over you and yours—you kill so that you and others, now and later, are not killed. I may be ashamed of that recognition, but from the millions of crawling and airborne enemies of agriculture, I know it to be true. They did not declare war on this family; but war their presence was, and our poverty their victory entailed.

They say farmers are good warriors because they learn to shoot at an early age. Sociological profiles done for the United States Army found rural, conservative boys from farms had a natural propensity to take orders, were in good physical shape, and naturally were accustomed to the outdoor life of nights on the ground. I suppose it is no accident that a Sergeant York was of rural stock. The Roman strategist Vegetius over sixteen centuries ago agreed. "I do not think," he wrote, "that there has ever been any question that rural people are the best equipped for military service."

At the heart of the rustic's ability to fight both effectively and brutally is this combative attitude toward nature—the very hardheaded view of the unending struggle to grow food. In my own family my Swedish grandfather who was gassed in the Argonne— my namesake, Victor Hanson, Jr.—who was blown apart on Okinawa a few weeks before the war ended, a cousin, Holt, with a bullet in his brain at Normandy, my late father who flew on the big B-29 thirty-four times over Tokyo, were all good warriors—the latter especially a lethal man with either a .50-caliber machine gun or 20-mm cannon. I think now that their skills derived from the years on this farm. These pest-sprayers and -dusters understood that you must—no, want to—kill things that destroy to survive, must obliterate those who would eradicate civilization, must as part of nature itself use muscle and brain to slay the slayer. When the warriors of this family recognized the character of the satanic Nazi or imperial German or racist Japanese, I'm afraid that not one of them blinked when his finger was on the trigger, and so slew many of those blindly obedient to an evil cause before being wounded, poisoned, and shot down themselves. Of this agrarian family's propensity this century to kill or die and be maimed overseas, my pacifistic, nurturing mother nevertheless once told me, "I think those Nazis would have killed every Jew on earth if somebody even as little as us didn't help to stop them." Many of our enemies, you object, were farmers themselves; and we in America did not fight to save the Jews, you add. Maybe, maybe not. But when it was over, the killers were dead and the dire consequences of fighting for fascism established—and those who fought the killers were free, and of a different mind and purpose than the killers themselves.

I now leave this catalogue of arthropods on a more mundane note, with a little-known but quite exceptional worm, one utterly unlike any other agricultural adversary and so deserving of some admiration historically accorded to the most formidable enemy. Unseen, stealthy, and alone of the arthropods, the nematode defies man. He stunts and then destroys orchards and vineyards with impunity. Like the submarine, the filthy worm operates in relative safety beneath the surface. There he cannot be detected or reached. Invisible and clandestine, his protective cover is the earth itself that deflects poison, be it injected, or gassed into his shapeless, dank subterranean domain. His method of attack is both surreptitious and ingenious. He crawls through the soil and feeds on the roots of the vine or tree, killing both imperceptibly. Their visibly dying leaves and shoots above are not always indicative that the evil is not from viruses or bugs within the stump, not from the bacteria of the air about, but from the unseen millions unceasingly chewing below. Only when the peach tree is uprooted or the vine bulldozed out do you understand the terrible things the lowly nematode has been up to those many months. The tree's or vine's chewed-up roots resemble a string of beads. Horrible galls swell up in reaction to the worm clan's biting. Root tumors choke off and block the transfer of nutrients and water to the condemned plant's leaves and fruit. These are natural cancers far uglier, far more lethal than any mutations spawned by man. Among us, as with nature, there are pathogens that gnaw at us so silently, so secretly, that they are known only when we at last topple and are no more.

Because the nematode feeds on and within the root system of the plant and beneath the surface, poison must be strong enough to soak and penetrate the dirt, yet not so resolute as to kill the food intake system of the plant itself. Few toxins can satisfy both prerequisites. There is, after all, damage, and then there is collateral damage.

Those nemacides that can do both, like the infamous DBCP, ultimately do more harm than the entire genus of this enemy, *Meloidogyne*. The wonder killer of the 1960s, DBCP, was cheap and toxic. With "wise use" it did not always destroy the uninfected roots of the tree or vine. It could, in other words, more often kill outright the nematode and not your vine. Unfortunately, the poison also did not break down after its task was through. Instead, after poisoning the nematode and saving the vine, it slowly leached well beneath the roots of trees and vines—down, down, down a hundred feet and more, on an unstoppable passage to Hell, polluting the very subterranean water table of a million humans above. In response, millions of dollars above went for new wells, filters, and bottled water to prevent Valley Man from becoming sterile as he drinks his DBCP-laced water. We millions are still paying for the nematode killer of the American farmer. The water I drank a minute ago is polluted so that this farm could have a grape crop forty years ago. Or so I wish to believe.

But who wants a productive vineyard when there is no one to inherit it? DBCP is in my well, my brother's well, my cousin's well, in every well around, they tell us. And they claim that even the shower and bath can bring it into our blood. The nemacide that lowers the sperm count of the male of the human species will always be with us. It will be forever young, long after it settles far below the world of nematodes that now thrive as never before— worms now free at last of their first and last lethal adversary.

Until that revelation of DBCP's traitorous conduct in the 1970s, I grew up knowing it by its trade name, Nemagone. It was a wondrous miracle drug that brought poor, centuries-old vines back into profitability, as they were at last given a much-needed, much-deserved reprieve from decades of incessant chewing. "Why," my uncle marveled, "with that Nemagone, some of my sandy vines are the best ones on the place; I've never seen anything like it."

He was right. No one had—then or since.

The worst thing that this family had ever done, even if unknowingly so, was to poison with a complex, unbreakable compound its wonderful, delicious water, its cool water that used to lie safe beneath our feet. Every man, if he is to live a full life, must create something that does not perish when he goes. For the farmer who does not paint or draw, does not teach, write, sculpt, or organize, his legacies are the sturdy sheds and irrigation pipes, the beautiful orchards and vines, the pure soil and water he leaves behind for others, the agrarian idea that he uses—not misuses—the earth to feed others. Of all the professions that exist, the American farmer's benefaction should not be the addition of a foul chemical to an otherwise pure subterranean aquifer.

What then in the post-DBCP age can the farmer do to battle the root-knot, root-lesion, and dagger nematodes? Without DBCP, not much.

He simply does not plant his vines in sandy, nematode-infested soil, where the porous passageways and the absence of alternative organic foods focus and amplify the pest's attack. A way to resist the nematode is simply to acknowledge defeat and farm apart from it.

He uses manure as fertilizer. Its worm-enticing odor and calories can, a few believe, divert the ravenous pest away from the vine's roots. Manure's mixture of hair and animal wools is also sticky, and can carve and cut even the smallest worm.

The farmer grafts his prize plums and apricots onto queer, feral rootstocks with names like Nemaguard and Marianna, wild primeval rootings that grow faster than they are chewed.

The farmer plants wild flora like lupine and sunflower between his vine and tree rows, whose roots make bitter the soil of the worm below.

He buys the latest chemical concoction, a poor bastard third-generation offspring of DBCP, without its ancestor's marvelous versatility—you could, after all, once pour DBCP into your irrigation water, drip it in with your cultivator, or shank it in the earth with your fertilizer. Most nemacides then after DBCP were always either too weak to kill the crawling pest or too dangerous for man to handle.

The farmer also fumigates his nematode-infested vineyard and orchard ground before planting. The early agrarians, with muscle and without money or technology, did so with water and tarps—either flooding their fields for weeks at a time to drown the worm colonies, or tarping the soil with black canvas to heat up and fry them below—both organic, though monumental, feats of logistics that rarely justified the exorbitant expense of labor and capital.

Now after the demise of DBCP, fumigation is most often accomplished before planting with the colorless, odorless methyl bromide, gassed in and then tarped over with plastic. It is the most lethal, dangerous gas in agriculture! Methyl bromide will turn your lungs into jelly in seconds. Methyl bromide dissipates into the heaven only to dilute the ozone. Methyl bromide, which every yeoman fears more than the worms beneath his feet. Methyl bromide, which allows the farmer to plant where cultivated plants usually cannot grow. Every time I popped open one of those gas canisters and tubed the invisible toxin into the soil, I wondered why and how I, a seemingly sane man, had purchased, but for a slight miscalculation, a horrendous instant death, chanced that my guts would become Jell-O. I wondered that, true; but thanks to methyl bromide, I also planted trees in nematode-infested ground.

Almost no farmer defeats the nematode. I say "almost," for some rather lush vineyards do now grow free of the pest, their roots unimpeded in their absorption of nutrient and water. How? Criminal-minded and sneaky viticulturists ship the banned and hated DBCP—which is a third-world godsend—up from Mexico. Those cheaters, whose tractors work surreptitiously at night and in the winter fog, prefer an extra half ton of raisins per acre, an additional two hundred boxes of plums, to a healthy sperm count of those water-drinkers in their midst. Are not the enemies of the enemies of agriculture at times enemies themselves to the farmer?

The nematode, then, reminds us of the Greeks' middle ground where the good life is found. In our rabid desire to free ourselves from all worry, all onslaught, living and inanimate alike, we ourselves create greater enemies than our enemies. When up against the nematode of life, be careful of DBCP.

The nematode has also an infamous counterpart: the ground-boring aphid *Phylloxera vitifoliae,* which does not eat vines, but rather entire vineyards. You may have learned of this gang. Phylloxera is the famous pest that sucked the vineyards of nineteenth-century France into oblivion and now does the same up north in the upper-crust Napa Valley. Fortunately for us, her haunt—all of the pests are what are called "oviparous parthenogenic females"—is beneath cooler soils. She is still too far north among the great wineries and boutique vineyards of upper California. I will be dead long before her slowly evolving, more adaptable offspring in mass finish their leisurely underground crawl south into our environs. For now, the present species of phylloxera would fry, not thrive, in the hot soils Fresno way.

Phylloxera, as is sometimes the case with these unseen enemies of agriculture, teaches us that each generation must fight the enemy in its midst, according to its own time, place, and station. We must, as these slow-moving aphids instruct us farmers, shrug and acknowledge that the foes on the horizon should properly belong to our children. We should not fight battles that are not yet ours. We should pass on our knowledge, show our scars, but under no circumstances believe that we can save our young and beloved by killing forever their enemies who have not yet arrived—who are still yet in cooler ground to the north.

Year in and year out, generation upon generation, the farmer's unknown war against the tiny, numerous, and unseen continues, violent, terrible, and unrealized by most other Americans in his midst. The lessons he learns from this unnoticed battlefield are more often than not both bitter and tragic: endurance above all, humility of course, but also suffering as the price of salvation. The farmer continues to struggle not because he thinks the battle can ever be won, but because he knows so well that on his watch, in his era, and according to his station, it must not be lost. And out of that apparent draw, he gains the wisdom that through such vigilance there comes a sort of victory after all.

LETTER THREE

THE FAMILIAR
ADVERSARIES IN
OUR MIDST

(Wisdom from Enemies That Are Not Always Enemies)

*For my part, I cannot blame them, but I should blame myself were
I peaceably to look on and let them carry all.*

—J. Hector St. John de Crèvecoeur,
Sketches of Eighteenth-Century America

The complexity of the more invisible foes of agriculture is rivaled
only by the diversity of its visible enemies. Unlike the microbes
and insects whose appetites are predictably relentless and instinc-
tual, those that we can easily touch and see are actually subtler adversaries, drawing us into a different sort of war in which the
battle is not always against the ugly, voracious, or unnoticed. The
greater clash, farming teaches us, is to fight the evil that does not
always appear altogether evil. The farmer learns from these other
enemies, the plants and mammals, that often the battle is ironic,
paradoxical, and not so clear-cut after all. Whereas he must always
fight without acclaim mildew and blight, and so brings that sense
of dutiful crusade to his community at large, against rabbits and

flowering plants, whose main crime is that they are not where they should be, the farmer learns that morality is not between absolute good and evil, but often a struggle only for something more good than bad. To keep our vineyards clean and our plums on the tree, we do things we often prefer not to, and in ways we do not like, because that is indeed far better than doing nothing at all.

I. Weeds

Like man on occasion, the enemies of agriculture can be cannibals of sorts: other plants that overrun the vineyard and orchard, strangling the producers of fruit so that they might grow instead their briars and thorns. There exist feral grasses and broadleafs that extinguish your trees and strangle your vines. These plant intruders rather like your artificially created habitat of plentiful water and food—now to be adapted for their own prolific, but largely unproductive, purposes. Designed to live in rock crevices, in drought, beneath the footsteps of thousands, without nitrogen or water, the weed indeed learns to prefer the milk and honey of the vineyard. There it comes to fruition in a way undreamed of by its genetic code. Weeds like to grow best where they were never intended!

Is there a lesson for us here as well, from the intrusion of this unwanted and rather uncivilized plant? There is, though it is brutal, illiberal, and simplistic. Food and abundance, superfluity and luxury even of shelter and sustenance, encourage the intruder, draw him into a world that is not his, but now one he surely enjoys. Bred and tempered for the wild, the weed rather prefers the cultivated, where his amazing powers of destruction can run amok among a weaker refinement.

I gazed for three years at an empty, ugly, dry five-acre plot on our farm, one uprooted of its trees and left fallow for want of a line of credit at the bank. Ironweed and puncture vine struggled to survive in the hot, empty sand. Sandbur found its natural environment scarcely adequate for proliferation. Thistle and wild willow—both species indigenous to that microcosm—became parched and died off. But on the March day when the new young trees finally went in, when the irrigation water went down the rows, when the fertilizer was shanked into the ground, and when the soil was cultivated and aerated, the weeds quickly threatened to overgrow their own once miserable habitat, choking the new plantings themselves. They rather liked what they were not, those seeds that lay dormant until someone else had made their life easy.

So too with some of us. We, who are the uncivilized and feral of humankind, hate our own miserly, ugly existence. We kill, rape, and maim, yet still tire of those kindred lawless in our savage midst. No, we are more eager to move across into an irrigated and fertilized field that we are not a part of. On a less nefarious note, we spray-paint our ramshackle fences, break the aging school's windows, and urinate on block walls. We torch the empty church, and deface the newly dedicated playground. We collect at the ATM, and loiter on the public bus. But we, who comprise the untamed, also demand at times a futuristic emergency room at the clinic, a new model of compact disc player, clean bathrooms at the local fast-food franchise, and anything we might enjoy but can neither create nor fathom nor maintain nor even in some small way contribute to. The wild among us, enemies of all culture, still rather like this late-twentieth-century civilized world—a now elegant, now rich, always relativist irrigated field to grow in for a while longer, to take from and overrun if need be. Those who were at home in the savagery of the wild like their transplanted existence among the well-nourished tolerant and tame, who say they are not all so wild anyway. The evil who move in suspect that there are others weaker—and more evil yet—who shall blink and let them in. Perhaps they fear only the few farmers left who have seen their kind in their own orchards.

Farming in California, I was once told by an octogenarian, is but "water and weeds." He meant you water, and weeds appear. He meant you must water and you must weed, and that is about all ye need to know. Weeds come in all shapes and sizes, with funny names like ironweed, fiddle-neck, bindweed, and lamb's-quarters, in addition to the suburbanite's Bermuda, crab-, and nut grass. But there is a general paradox of what is and is not a weed. On occasion, a weed is an unwanted plant, not intrinsically foul per se, merely one not welcome at a given place at a given time—its leaves and stems are not always thorny, its flowers not exclusively small, its fruits only occasionally bitter, poisonous, or nonexistent. In your yard morning glory and four o'clocks can offer visual relief. In your vineyard they are sinister intruders that throttle your vines. As tomatoes feed man, so their feral cousin, deadly nightshade, seeks to kill man or impede his work. As hollyhock adds color to man's walkway, as mallow feeds the poor in his times of famine and poverty, so their kin cheeseweed overruns orchards, destroying, not providing, food. My poor grandfather in his eighties, the old and tired veteran of seventy years of the weed wars, finally used to spray herbicide on the lawn, shrubs, and ornamentals of his own yard. When pressed, he remarked, "They're weeds; they don't have fruit." As his brain shed the subtleties of culture, the elemental instinct, if now warped with age, nevertheless remained strong—be on the guard always against those, even if beautiful and fragrant, who would extinguish the food makers, who would join, mingle among, but ultimately drown out the bringers of harvests.

Almost all weeds are feeble enemies of agriculture should they grow where you wish: that is, in the world of arboriculture and viticulture, between the rows of the trees and vines. In the modern age of the vineyard tractor, trees and vines are planted in grids anywhere from twelve to twenty feet apart, to accommodate the machine, an ample killing field for weeds, which sprout in the well-watered furrows. Five acres of vineyard can be disked each hour by the vineyard tractor and cultivator, an entire generation of weeds mulched to add its essence to soil, thousands of plant carcasses to be absorbed by tree and vine.

The only caveat? Disk the brood down early when they do not clog the disks. Miss a July week or two and weeds grow tall, far above your whirling blades. Then the bouncing and skipping cultivator is not unlike its driver himself who struggles to shave his four-day-old beard, the stubble clogging and dulling the razor that nicks and cuts his flesh as much as the hair itself. Do not trust the weed to stop growing on its own, or to leave enough water and nitrogen for the softer, tamer tree. It won't. Like the ravenous mite, it grows and thus kills until stopped.

Most of these weeds, remember, are not dainty flowers, but tough perennial bushes, with roots half as deep as the vine itself. Weeds, I think, are all those who live for their roots, rather than roots for their fruits. The worst, Johnson grass, cannot be pulled out by any normal man. To dig it is to separate and thus spread its infectious rootlike nodules. To spray it is to retard, but not halt, its growth. No, to eradicate this noxious perennial weed, you must chemically sterilize the soil, spray the leaves, dig the roots, and gradually turn the environment against the plant itself. Curse the rancher who introduced that foul foreign species into California, the would-be-ubiquitous cattle grass that ended up beneath the vine to steal and siphon off its food and water. Give Johnson grass and a vine an equal playing field, and the weed wins every time. It was meant to.

So if you let weeds sprout beside the vines and trees on the berm, and not in the middle of the rows, their infestation becomes lethal. Beneath vine stakes and wire and drooping canes, the weeds infiltrate into the farmer's safe zone of culture and cannot be reached. Beside the vine they grow up indistinguishable among leaves, canes, and grapes—hiding, stealing water and fertilizer, chafing the picker, leaving seeds and prickly residue amid the fruit, allowing its own pests to hop over to the surrounding tastier vine leaves. Weeds wish to get as close to the vine as they can. They know the protective zone of refinement, and they surely like it. They may not be mobile in the animal sense; but once established, weeds are the greater foe—as plants they are more a part of the earth than the animal, harder to kill outright. Smash a bug, it at least dies instantaneously; pull out a weed, and it sort of dies, depending on where you toss its root—or whether you get all of its root to begin with, or whether it has already seeded out.

As in life, so in agriculture, you must keep your berm of culture free! There are only three ways to accomplish that task—and timing is critical with each. Spray the berm in winter with cancer-causing, water-table-polluting "preemergent herbicides," which sterilize the soil beside the stump for a year. The trick is to get the right dose at the right time: too little poison and the weeds grow anyway; too much and the vine sickens; too early and the toxins tire prematurely; too late and the weeds have established too strong a beachhead.

Choice two is the "postemergent" species of plant toxin: "knockdown," or systematic, herbicides that either burn up the plant or infect its roots "systematically," shortcircuiting its biophysiology itself. But the problem here is tricky as well: "knockdowns" kill the visible plant, but not its roots, and so they sprout right back, stronger for the setback. Systemic chemicals do indeed kill the roots. But they sicken your vines as well if you miss with a few droplets and hit the canes, bunches, or vine grape leaves.

A whole science exists to spot "systematic herbicide injury" in trees and vines. A whole subspecies of Valley lawyer thrives to sue those who accidentally spray what they should not. Most farmers, as would-be chemical strategists, mix knockdowns in with their winter preemergents, come back later with systemics, and so use three types of herbicides to keep their berms free—even as the chemical companies brag that a single "all-in-one," "single punch" strategy is quite sufficient.

Choice three of weed eradication is clean and exploitative alike—the back of the man who needs the minimum wage. The dirt shoveler and callused hoer of the ages walks hundreds of miles in the vineyard, up and down your rows, a lifetime of digging out ironweed, orchard grass, lamb's-quarters, barnyard grass, filaree, fiddle-neck, sand- and cocklebur, and puncture vine too, pitching them all out into the middle of the row to fry in the sun and be disked up by the tractor.

You object that in our modern world of hydraulics, lasers, and computers, machines can—and should—be devised to cultivate the vines within the row. At the millennium we should have figured out how to take out the weeds and not the stumps, eliminating the dust, poison, and human misery of removing weeds that grow up on the vine row itself. So said they all, even at the turn of the century when they introduced the Kirpy plough, the French plough, and then in the next half century the improved "Weed Badger," "Weed Eater," and "Weed Blade," and now again the futuristic "berm sweeper," "berm tiller," and "berminator." These automatic retracting cutters all work—for a while. But the quagmire of vine wire, stake, and stump eventually prove too much.

But unsolved by the machine is the fatal contradiction now well known to you in the battle against the enemies of agriculture: that perfect but unattainable mean between power and grace, between force and restraint, between protecting the good and killing the bad. The machine's blades, knives, ploughs, brooms, and teeth must be hardy enough to uproot the ironlike roots of perennial weeds on the berm, yet amply nimble and subtle as the machine moves down the row to jump suddenly back free of the upcoming grape stake and stump. The in-row annihilator must kill everything in the six-foot zone between the vines, but under no circumstance become sap-drunk and continue its mayhem an extra inch onto the vine and its stake.

Most contraptions do not stand up to that in-and-out, hit-and-retract wear and tear of uprooting the weeds—but only the weeds—between some two hundred vines down a quarter-mile row. Those more rugged in-row cultivators that can uproot the target and do it consistently have a bothersome tendency to be a little too eager or too clumsy. In each row they take out a vine along with the weeds. A hundred rows every thirty acres or so, and the dead vines per row add up. And I say nothing of the eyes and neck of the man on the tractor who must guide the "labor-saving" machine. Is it not a contradiction that one must look forward to drive the tractor while looking back to guide the cultivator between the vines? No wonder my brother had to have his eye muscle trimmed and cut; a man who looks over his neck for forty hours a week begins to look over his neck the other ninety-odd as well.

May I also pass over this decade's lasers and electro-shockers, which purportedly incinerate every weed that does not conform to their computerized memory banks? I leave it to you, reader, to evaluate the utility and durability of a device that uses cameras, computers, and electricity in a filthy vineyard world of dirt, mud, and grease.

So to control weeds, you pollute with the tractor and disk. The engine spews its little bit of blue-gray diesel exhaust into the already dirty Valley air. Its trailing disk that rips the soil contributes a cloud of dust and particulates to add a natural component to the smog. Alternatively, with cleaner chemicals you can poison the soil of your vineyard, soaking its earth, coating the stakes and stumps with some of man's most vile creations: 2, 4-D, Paraquat, MCPA, diuron, simazine, and arsenic. Or, alas, you can hire the age-old man with a hoe to dig and hack as he swelters and aches beneath the 100-degree Valley sun—a man in dire need of a union to demand what he sorely deserves from the bankrupt farmer. You in the city have forgotten that the entire cosmos of DDT and diesel smoke was designed to save the back, the arms—the life—of the man who could not read.

The war against the weeds can drive even the strongest to insanity. I have seen some of the cleanest farms of this Valley fall to bankruptcy. When prices dive, the weed fighters are loathe to cut back on their chemicals, implements, and hired diggers. No, they would rather go broke than leave their berms dirty—as if the weeds, now free, might take over and strangle the world itself. And broke they go. Last week one phoned me: "You know the old McGill place? Well, that was mine until it was auctioned off. Not a weed on it either. Now those new guys have left it a mess."

Mr. McGill never made the fated connection between his diligence in keeping his orchards and vineyards spotless and weed-free, and not cutting his costs when prices fell. But then Mr. McGill was a weed fighter. He was the good farmer in this cosmic fight against disorder and the uncouth, who cared little whether raisins were $500 or $1,300 a ton, whether plums sold for $4 or

$14 a box—if his berms were weed-free. And so Mr. McGill is broke and the McGill farm is no more.

The weeds of this world rather like the cultivated environment they cannot create. They teach us that the cultivating arts can make us soft even as they save us, that knowledge can cloud our moral sense even as it enlightens us, that our progress can allure the rugged into our midst—those who are not quite as civilized as we would prefer. We need these rare weed fighters and their blinkered devotion that seems to us so uncouth! I don't think I ever understood the hysteria of late Roman imperial literature and its constant worry and paranoia over frontier defense until I returned here to farm and join the fight to cultivate. When I left the university, I once used to lecture my grandfather on why he should "let the weeds go" and prune his costs—only to be taught the unpleasant relationship between intellectual progress and moral regress from a suddenly rank weedy vineyard north of the house.

We farmers, it is true, use very little that is pleasant to be rid of the weeds from our presence. So we must create enemies for our enemies that are nearly as foul—and most often more ridiculous. Keep the herbicides in the barn, and the weeds have the vineyard. Make the ground too sterile for the unwanted plant, and the ground, though free of Johnson grass and lamb's-quarters, is never again as it was and should be. It is a zero-sum game, where the price and diligence to defeat the weed is not cheap, and can take us down as easily as the weed itself. I think in America now to keep the local park—tree-lined, studded with swings and tables, well lighted, with water, clean rest rooms, and sanitary trash receptacles—free of the gang-banger and molester, clear of graffiti, feces, and syringes, our patrollers, police dogs, and cleanup crews have to make it almost no park at all.

II. Trees and Vines

How might fruit trees and vines—the archetypes of civilization itself—be enemies of trees and vines? Remember that the farmer
represents culture, not nature, and his planting stock, in contrast,
is natural, not entirely man-made. True, man breeds and cross-
breeds his trees and vines. But his raw material of creation is cellu-
lose, not plastic, and it is nearer to the wild than it is to man. An
enemy of agriculture, no small one at that, is thus the cultivated
plant itself.

The tree and vine without the farmer, as Virgil knew, becomes
no tree and vine. Should the yeoman not prune his trees, thin his
bunches, sucker his stumps, or pull off his leaves, his cultivated
trees and vines will go back to the netherworld of the half wild.
The vineyard and orchard, like the stallion and ox, do not wish to
be completely broken, to be ridden or hitched always when they
should roam. Miss just one season of pruning, do not sucker one
year of stump growth, decide you are tired of pulling off excess
shoots each spring, and your farm is nearly unrecognizable as its
stock heeds the call of the savage. Two seasons of such neglect and
your trees and vines are feral. Their growth becomes woody and
rank. Their leaves turn diseased; even their shapes are unrecognizable. Those grafted trees and vines once left untended are no
match for their own wild rootstocks, which quickly send out ugly,
fast-growing shoots to dethrone their weakly, more refined masters. The untended are more now a part of the world on the ditch
bank and abandoned city lot than of your ordered lines of cultured and well-mannered food-producers. They are thorned and
they are ugly, and they belong caged inside the trunk below the
graft, where their exuberance can only push and feed, but not
overthrow, the weaker but more useful limbs.

I know this cycle from experience: one winter I let all the suckers grow on young Mandarin oranges, rank, deep green, spiny shoots that in but a season smothered the refined and pleasant limbs emanating from the domesticated graft above. The orchard was gone to the wild, but it took a bulldozer to dig out those stumps, once the alien wild root took control of those poor orange trees' weaker and tamer genes. The sheer terror, and public embarrassment, of that orchard has not left me. Those poor trees were suffocated by their own wild thorns and roots. I learned what a single winter of lawlessness can do, what a surrender to our own fury becomes.

This balance too between the natural and the cultural, this *metron,* as the Greeks knew, holds enormous wisdom for the cultivator. It confirms that within us all, beneath the veneer of libraries, heart bypasses, laundered clothing, and eyeglasses is the wild. The wild will tear and chew within us without the bridles of the *polis.* Man, like his trees and vines, is not at birth a kind, benevolent, or humane creature. He is not by nature a pretty nice guy. Take away custom, religion, and law—take away his graft— and he is just as liable to eat, torture, or rape his brethren. He will use his teeth and employ his nails, and his muscle will not serve his intellect. So, the spiny wild-root stock of the plum, the exuberant underbrush of the peach, the useless myriad of tiny second-growth grape bunches are like the passions and appetites of man himself, which must at all costs be pruned, suckered, and thinned if he is to remain civilized man—and not become Hamlet's "unweeded garden, / That grows to seed; things rank and gross in nature / possess it merely."

Arboriculture and viticulture teach the farmer that the beautiful and powerful life force is, after all, wild, beautifully and powerfully not nice. Often the wild root is stronger, more impressive in its dash and bold honesty than the cultivated. It even holds a dangerous and perverse attraction for the pampered and the tended. The polished writer, the smooth-handed academic, poet, and actor are especially prone to its allure, and sometimes wish to dabble safely with the weed. Be careful, you refined shoots above the graft, of your wild, strong, alluring—and savage—roots below. The farmer, as civilization itself, stands foremost, must stand preeminently, between nature and this abyss. The agrarian must view his neighbors, his own kin—himself—with caution, should his pruning and cultivation fail. The graft of refinement onto raw energy is fragile, and it needs our constant surveillance, our steady work, so that it will not break and let our root grow free. And thus we all really do need the farmer, alas, to warn us of the peril of a powerful and beautiful and often not nice natural world that is everywhere—and in us.

III. Mammals and Birds

There is a melancholy inherent in killing those higher up on the food chain. The winged, legged, feathered, and haired bleed red and make queer, but noble, noises *in extremis*. No farmer enjoys their death, be it by poison or gun. No farmer likes the warm-blooded of his farm to bleed on his soil. Every farmer dislikes killing the fauna for a variety of very understandable reasons: the visible and animate enemy is now disturbingly closer to his own species, its size, appearance, and neurophysiology more akin to the farmer's own. It has fur not scales, lungs not pores. And so are they really enemies at all?

At twelve I once stepped on the neck of a wounded rabbit that I had shot from too far a distance; thirty years have passed since it shrieked as its windpipe was crushed beneath my tennis shoe. It ate our young vines, it is true. But I can hear those near-human sounds still this week. Rabbits do cry like men when dying. The spot of its demise on this farm I avoid today and will tomorrow, three decades later. There are, I learned from that woeful crushed bunny, enemies we kill that we should not. We should not write poetry, as did the Greeks in their anthology, about the extermination of the lowly hedgehog, "Its body full of sharp spines, a grape-eater, the destroyer of sweet vineyards, caught all curled into a ball rolling among the grapes." We should not step on the windpipes of rabbits that are dying slowly from poor marksmanship and too few bullets.

Killing the warm-blooded, then, is different from gassing the arthropod or digging the weed. They may, as some of the Greeks thought, have souls; Montaigne thought they had both "choice and industry." In contrast, there is no society for the prevention of cruelty to plants, no statutes punishing the torturer of insects. Worse, the damage done by birds and mammals, reptiles as well, is minuscule in the great hierarchy of disasters wrought by the enemies of agriculture. The tooth and claw of a few is not comparable to the arthropodic mandibles of millions. Does the farmer, then, ignore, as he should, the young grape rootings eaten by the fuzzy cottontail, the almonds now stored in the burrow of the cute and so-hungry shadow-tailed squirrel, the grape pecks inflicted by the famished but proverbially wise and iridescent raven or the curled-up hedgehog?

He does not!

The yeoman instead buys nets, loud cannons that simulate howitzers, and plastic hawk kites. He stocks up on warfarin and strychnine that thin the mammalian blood to the point of stroke and cerebral hemorrhage and ulceration of the stomach. He orders .22 long-rifle hollow-point ammunition by the box of five hundred. The agrarian has his shotgun loaded behind his pick-up seat, a case of gas canisters and tubing in his truck bed on reserve. No, the farmer attempts quite viciously to eradicate the squirrel, rabbit, and starling, whatever their alluring hue, their kind visage, their admirable efforts at cohesive and brotherly organization.

Why does he do so? To protect his kin from bite and claw? To ensure food is on the table? To meet his loan at the bank? Might perhaps these rodents carry disease? Is rabies or plague a danger among the warm-blooded living within the trees and vines?

Rarely, if at all.

The farmer kills the larger enemies of agriculture—visible, beautiful, and of vanishing number—because they are there—there where they should not be. And he can kill them when he likes. They do, after all, represent the idea of intrusion and challenge, in a most arrogant and visible manner. They apparently assume the vineyard or orchard is still the burrow or hollow of a century past. Animals insanely presuppose that because their kind has had the land for a million years, yours but a hundred, it should therefore belong to them. They, like the farmer, have an elemental will to live, to kill even to preserve themselves. On this planet they should know and accept the wages of that unending scrap to eat. In theory, they are a most brazen pest!

Mr. Crèvecoeur realized the irony of it all.

Such is the nature of man's labours and that of the grain he lives on that he is obliged to declare war against every ancient inhabitant of the country. Strange state of things! First by trials, by fraud, by a thousand artifices he drives away the ancient inhabitants. Then he is obliged to hunt the bear, wolf, and fox.

"There are squirrel holes all over that bank by the Santa Rosa plums," I once complained to my twelve-year-old son. "Now you get out the .22 with the scope, go down over there, and lie behind the walnut tree and shoot as many as you can," I ordered. (My grandfather, after all, when I near tears had brought back the dead bunny with the crushed windpipe, had said—before he skinned and ate it—"If you can get a few more like it, those young vines may make it yet." I did not, and the vines, of course, made it.)

In their famine a few of the more desperate and hungry squirrels had actually climbed up one of the adjoining plum trees—a single one of three hundred in the orchard—and eaten its twigs bare. Such insolence, such impudence, such a serious fifty-dollar loss! Worried that my son might kill but four of the colony's thirty and more, I plan to drive down with the tractor and pipe its noxious diesel exhaust into their holes. Worried still over the impotence of his missiles and my monoxide gas, I will add strychnine-laced grain and raisins to the runs (my brother, I later found, beat me to it). Still, I wonder just how dangerous it would be to drive the tractor right on top of that hill, its one-ton disk and three-ton powerhouse tottering as it grinds up those burrows. Yes, I do weigh the danger of being crushed as the tractor totters and flips me over the edge of that bank, versus the annihilation of the entire brood of single-tree-eating squirrels. I forgot entirely what Mr. Crèvecoeur said of the war against the squirrel, "As we pay no tithes in this country, I think we should be a little more generous than we are to the brute creation." Perhaps if the local, state, and federal government would cease taxation for a year, the farmer might give the squirrel burrow a reprieve.

Now, reader, upset over the killing of the furry and tiny, you will interject that our war against animals also offers man too many negative lessons: we are vindictive, cowardly in attacking what we can see and hear but what is otherwise calm, defenseless, and of apparent, rather than real, danger. All this is true. We are.

In our Valley, just as its farmers shoot birds, poison rodents, hunt down (illegally) coyotes that do not ruin their farms, so too the sheriff, the county agent, the federal regulator hunt out the yeoman who is not a challenge to the law. But like the rabbit and squirrel, the farmer is simply there to be had, the self-employed relic that causes no real harm, the last man alive of this Valley who listens, obeys, and pays up. In Fresno County, thousands drive their unregistered cars inebriated; thousands more con their welfare checks and sell their food stamps; hundreds loot, stab, and shoot; a select few bribe and pay off to turn vineyards into Vineyard Acre Estates; and we do little. Like the farmer's own war against the virus and bacterium, the government enforcer does little because he can do little.

But the farmer knows all this and carries that burden with him to such a degree that it becomes a part of him. When he is ticketed for exceeding the speed limit by three miles while felons speed by at ninety, he knows why. When he is fined for not burning his chemical bags, while the company that made them is allowed to dump their detritus in the ocean bay, he knows why. He understands that justice is capricious and often poorly meted out; that its enforcers are often slothful and opportunistic; that the visible misdemeanor is easier for the cowardly to address than the unseen felony. The farmer realizes the inequity of justice and yet concludes that it is foolish to seek refuge in the idea that punishment for his petite crimes are not justice. The farmer, almost alone now in society, knows that he was exceeding the speed limit and thus was a menace to his peers on the road; that he must dispose of his bags if the wind is not to blow their toxic residues into the air. He knows all this because he too guns down the industrious and attractive squirrel in his plum tree, even as he cannot prevent blight and scale from taking his entire crop.

IV. Weather

Let me regress for a moment from the struggle against plants and the larger animals to chronicle a force that is both evil and good. Nature, of course, in the form of volcanic eruption or earthquake can dwarf all the enemies of agriculture combined. It can pollute the air far better than the cars of Los Angeles, flatten cities more efficiently than any bomb from that queer brain of old Dr. Teller, clear-cut lumber quicker than any corporate raider now turned desperate log company CEO. But thank God in our Valley that the worst manifestations of nature—shaking ground, air convulsion, primordial ice—are distant. Less lethal manifestations of rain, frost, and drought are instead what we fear. While they do not often kill us, much less level our vineyards and orchards, natural calamity can cleanly, almost surgically, take an entire crop in a few seconds. Such intemperances are rarer than the living enemies. But when nature as adversary visits, it leaves disaster in its midst.

An April frost turns an entire vineyard black in hours. A March hail shreds every peach on twenty acres. Six years of drought take a decade off the life of an apricot orchard, and leave your wells dry, your pumps sputtering air. A big wind in May can blow off your entire plum crop, just a day after you paid $20,000 to have it so carefully thinned. A rain at harvest? It can do what a million worms, a billion spores cannot. You can go to bed with 100,000 trays of beautiful drying grapes, and wake up to stinking mush when the big hurricanes blow in from Mexico. So much for the money that was saved for blacktopping the dusty driveway this decade—that anticipated capital is now rotting out in the vine-yard. Yes, Mr. Crèvecoeur, man struggles and survives.

The weather teaches the farmer that he is to the elements as in-sects are to him, as old King Lear's "flies to wanton boys," as tiny, as every bit as fragile and inconsequential when something big on the outside decides he is all through. The bitten farmer is often surprised at temperate weather and a kind nature; often in those rare years he mutters, "My God, it did not rain, hail, or freeze this year for some reason."

Weather leaves the farmer dumbstruck, silent, blaming no one. Plato, who knew of "the fears of the farmers," wrote that "one bad season makes the farmer helpless." Natural disaster, I think, re-minds the agrarian that he is but man in a rather large cosmos, re-news his knowledge that nothing is properly his—ever. Long ago, in April of 1971, my eighty-one-year-old grandfather told us, "Tonight we are going to lose every grape bunch on this ranch, and there's not a thing we can do about it. Oh, I'll run the pumps and the ground is sealed up pretty good, but there's a big cold from Alaska and it's going to be twenty-four by midnight." A man who despised the local motorcyclist who roared down his or-chard, and who hunted down the horseback rider manuring his alleyways, laughed when his vines turned black and he lost his year: "I guess we have no business thinking we're in control when we aren't. A couple of more years like this one, and all those peo-ple down in L.A. would know where their food came from."

If you are going to lose everything in this world, is it not prefer-able to be robbed by water, ice, or heat, than by an ugly worm, an invisible bacterium, or a hooded assailant? At least, I learned from this ageless enemy, natural catastrophe means you no harm when you are in its way. It has no particular look, no static shape. Rain and hail and frost are simply there, with you, are you—and cannot be sprayed or cut. It is far better to fall before something greater, something spontaneous, something divine, than before the more ubiquitous smaller and premeditated predator. The unexpected rainstorm, the unpredicted frost, the unforeseen hail—forgive me, farmers—are a catharsis, a natural cleansing that even the ruined acknowledge. No farmer swears at God when his vines are ruined by frost; all farmers swear at man when ten vines are taken out when the errant drunk driver swerves off the asphalt.

Bad weather is not always bad, if it reminds you that you, like all of its foes, are nothing. Instead, the absence of natural challenge, the presence of fertility and temperate weather, is more deleterious in the long run—the Greeks would say worse—than poor ground and a harsh climate. Nature the benefactor is often no benefactor. When the soil is rich, the days calm, the water plentiful, man grows fat and docile, works less to survive, assumes more than he should. He can get three tons of raisins an acre with half the fertilizer, water, and cultivation than the poor wretch on the sand hill, who works dusk to dawn to eke out his ton and a half.

Again, is it surprising that Herodotus located the national character that won at Marathon and Salamis within the rocks of Attica itself, and the bounty of Asia as the culprit for the Persians' softness and subsequent defeat? Do you fault Dr. Hippocrates for believing that the Greeks outdoors in isolated valleys and rugged hills were a more combative, more healthy and free folk than those in the more fertile hydraulic dynasties of the East? Is it not true that those who farmed in New England among ice and boulders were more than a match even for the brave and martial denizens of the South's deep loam?

I have seen the farmer on inorganic sand embody the nobility of that struggle in a way unknown to his wealthier counterpart on heavy white ash ground just a few miles distant. Just as the Valley resident who endures the furnace of continuous 100-degree summer days is of another world from the temperate Marin County cool suburbanite. Just as the yeoman on his thin earth is not of the same world as the local agribusiness magnate perched atop black bottomland, so nature can sap, weaken, and ruin us through insidious kindness, through the vast production of harvests that are unwanted and unneeded—once it is altogether no longer an enemy of agriculture. More farmers have gone broke producing bumper crops than losing half their harvests to frost. Is it a cruel or benevolent nature that gives us our leisure and affluence? Is it a capricious or an all-wise storm that sometimes takes our crops?

Part Two

MAN VERSUS MAN

THE HUMAN KIND

(Wisdom from Enemies Who May Appear as Friends)

The most ignorant, the most bungling member of that profession will, if placed in the most obscure part of the country, promote litigiousness and amass more wealth without labour than the most opulent farmer with all his toils. They have so dexterously interwoven their doctrines and quirks with the laws of the land, or rather they are become so necessary an evil in our present constitutions, that it seems unavoidable and past all remedy. What a pity that our forefathers, who happily extinguished so many fatal customs and expunged from their new government so many errors and abuses, both religious and civil, did not also prevent the introduction of a set of men so dangerous.

—J. Hector St. John de Crèvecoeur,
Sketches of Eighteenth-Century America

Sometimes survival requires brute strength and unending struggle against nearly invisible forces that are cosmic and impediments to civilized life itself; on other occasions, the farmer is confused as to what is good and what bad in the natural world, as a plethora of flora and fauna intrudes on his space and attempts to live at his expense. But even more challenging is man himself, who sometimes presents himself to the farmer, like the bacterium and nematode, as invisible, insidious, and relentless, at other times like the rabbit and bird as a bothersome nibbler who must nevertheless be reluctantly removed from the vineyard. But worst of all, the farmer learns from his human brethren in agriculture that great evil sometimes arises from the better of his species. Good men, because they are mostly good, are enemies hard to fathom and even more difficult to battle with; while the bad, it turns out, are not so bad after all because their aims are clear and their dark work manifestly dark. Anyone who has farmed can teach those in town that they must fear their good friends and those who wish to be their good friends sometimes as much as their enemies.

I. Bad

Farmers, like their kindred in town, as a rule regard all of their own species whom they do not know, but who enter their domain, as enemies—Hobbes's world where exists "a condition of war of everyone against everyone." The ditch-tender and mosquito-abatement official alone he lets off with a gruff "about done?" His industrious neighbor (who is a mirror image of himself) he watches carefully: boundaries, easements, and common rights of usage are the fuses that ignite century-old feuds between those who can go nowhere else. So you chafe at, but under no circumstance do you strike, your neighbor.

In general, like the rodent and crow, even the human trespasser does no real damage. But again like the mammal and bird, man the intruder, who enters the farm where he should not, offers only an affront to the farmer. His is the insolent challenge that he can do what he wants, go where he wishes, on land that is not his own, this man who has no concept of "own" at all. But quite unlike the animal, who is without speech and possesses a nobility from his dumb existence, the agrarian bounder has no redeeming virtue. His reason, voice, and common humanity usually reveal him instantly for what he is: a breathing, animate no-good, without excuse, worthy of no pardon. The farmer feels no remorse in blocking that thoughtless and selfish encroachment. No wonder the ancient Greeks put sharpened sticks among their grapes to spike the interloper. The trespasser somehow has come to embody the entire contagion of late-twentieth-century life in America: disrespect for property, an attitude that rules are not binding, insolence rather than appreciation for the leisure and affluence in his midst. He surely is the evil incarnate that tiny Mr. Madison worried about when he drew up the Constitution.

Fifty and more I have escorted off this farm, the more so now as the housing developments are but a mile away. The benign in summer are often female. Many are young. A few are often scantily dressed. They usually approach on horseback, riding arrogantly on black steeds through your vineyard or trees.

"Hey, dude, just tell me where else I can go," they whine when confronted—as your eye catches a tiny skull tattoo on their bare midriffs, "Metallica," a Swastika or 666 brand burned into their horses.

The mostly male and adolescent, even when not entirely felonious, are far worse. They think your rural world is but a big funhouse. To them your orchard is a rather large circus in which to drink, fornicate, shoot, and litter as they please. "Chill out, farmer man, we're just kicking back out here," they insolently proffer.

But they lie. Most are thieves—petty thieves, but thieves nonetheless, who are there to take and profit at your expense. The more determined of them customarily work at night in town, when their prey is asleep and traffic scant. But occasionally, when low on drugs and when gang attire has gone out of fashion, they venture out onto the farm for fresh capital as day-stealers. "The land is so big, the people so few," this species of dimwit must reason, "what chance is there that a single man will see me at work?" But the farmer spots them from afar as he is digging weeds, on the tractor, or tying vines, and he takes off at a run toward the nearest.

"OK, mister farmer," the criminal begs off, "I guess you caught me red-handed, but I thought all this iron didn't belong to nobody anyway. Why do you need this ole rusty stuff anyway? What you gonna do, man, shoot me over a few dollars of scrap metal? Come on, cut me some slack."

And of course we always do, when mostly we should not.

Ultimately the farm can attract the psychopath. This is the rather rare but calculating villain who reasons that naive farmers' doors are unlocked, that women can be in the house alone, that his purported flat tire or an empty gas tank might win entry into the hayseed's parlor to "use the phone," that where no street light or police car cramps his prowl, it is but his muscle and gun barrel against yours. The true criminal's ultimate goal is unknown even to himself; if he is not casing your farm for booty, he is scouting out your barn and shed for potential meth labs, chop shops, or drug havens. If on arrival there are tractor batteries and chemicals about, he will steal. If there are women alone, he will rape. If the gas tank is unlocked, he will burn. If caught entering the bedroom window, he will shoot. He reasons only that your farm is like the frontier, the marshall a day's ride away—and the spoils go to the more audacious and vile.

But unknown to the coward, the farmer himself often rather likes this sudden universe of do-it-yourself justice, now hand-delivered to him on a platter. For once in his life—unfortunately, if he is not careful, maybe for the last time in his life—the old weed-fighter's and bug-killer's own natural world is free from the county zoner and state inspector, and more dependent on his own nerve and muscle.

To the farmer who rises at dawn, who beds early, who works when he should not, man, the gratuitous criminal and predator who would kill or maim, is his most despicable, though not his most serious, adversary. It requires all the farmer's judgment, his restraint, his thoughts of his wife and children, not to pull the trigger when it seemingly, though wrongly, would have been better for all around that he had. I suppose when confronted with the enemy that can no longer do us the harm it intended, it is better to kick it away than kill it and have its stinky carcass on our porch, on our hands, on our minds—on our souls.

Yet I acknowledge too that the more we find material comfort, ease, and knowledge, the harder it is for us to rise to the moral but very mundane and dirty task of identifying and combating human evil—evil that so often stains and warps those purer who must keep it at bay—evil that arrogantly demands for its own extinction all that we value and have worked for, sometimes our own clean, private, and moral lives if need be. Pericles did say that courage was found only in the sacrifices of those who had something to lose in the first place. It is so much easier, cleaner on the mind and soul, to worry for a well-publicized moment over the newly contrite, often religious, and publicized condemned killer on death row than to fret over the messy and bothersome scene of the orphan and widow—much less the long-ago murdered victim rotting away beneath the cemetery grass, long forgotten by the society he once served.

II. Good

Society, complex society I mean, has become the final, the most unknowing, and the most lethal enemy of agriculture. This ultimate enemy in the lake of fire appears as the most numerous, the most visible and nicest pest around—one that cannot be sprayed, plowed, or uprooted.

But who is this "enemy"? The good, law-abiding, and hard-working suburbanite in the $30,000 Grand Cherokee who makes it a point to complain about the $3 watermelons at Wal-Mart? The good and philanthropic Boston hostess who rightly worries over chemical residues on those large, bright yellow May 1 apricots from Fresno, flown 3,000 miles to arrive just in time for her garden reception? The good, earnest, and kind business major who is offered an entry-level $60,000 salary to devise joint-venture investment plans for repossessed farmland? The studious and good kid who plops down a dollar for a Snickers, but is agape that a red pear should cost 50 cents?

Perhaps all of them. Perhaps none.

The ultimate enemy is everywhere and nowhere. They are not the bad potbellied, mustached foreclosure officers of agrarian lore. They are not bad, blow-dried, white-collar criminal brokers who stealthily overdeduct the farmer's check, who brag of all the raisin disasters of the last half century—the frost of 1972, the rains of 1976, 1978, and 1982—when men lost their crops, their farms, their lives, as the Golden Years, when they profited as never before. They are not the bad, greedy developers of myth who want six houses for every five hundred vines. At least not entirely they are not.

The ultimate enemies of agriculture are more insidious and imperceptible. They, like you, are actually rather nice to see and meet. They are ourselves: "good people." But they, who work so hard and so long at hospital, plant, and office, have become—have had to become—accustomed to cheap food, to the economy of scale at all costs. They want food pretty, cheap, and now! Always. And from very far away! Whatever the cost, damn the consequences. Faced with their own drudgery, they must rely on what the farmer gives away for free, his food, his land itself—himself— to tide them over. In short, with VCRs needing repair, with mortgages past due, kids' braces that did not take "just two years," and the water pump out on the Honda, they must expect—and can always get—food at the only price they are willing or able to pay. It is true of all of us. Because our food is so inexpensive, so attractive, safe, and plentiful, they have a margin to put our money elsewhere.

So in our society, it is the good who are such a silent, sturdy, and formidable enemy of agriculture. No insect ever has done as much damage to farmers as a colony of well-spoken and usually quite friendly food brokers, who shipped pears, plums, peaches, and grapes off this farm, year in and year out, in the 1980s for $5 to $8 a lug to me, $60 to $80 to the consumer. I once saw a plum box from our farm in Manhattan priced at about $85 a box. Shoppers were picking Santa Rosa plums out of the very box I had seen a few hours earlier in our packing shed 3,000 miles away, I believed the very box we received $6 for.

The 1993 and 1994 return for the "reserve tonnage" of our raisin crop held for two years by packers was to be paid out at 20 cents on the dollar. We would have been far better off to have lost the entire year to a March frost, than to grow, pick, and truck in hundreds of 1,000-pound bins to be sold for the price of cattle feed. But good brokers, who exchange Little League stories, who show you their office pictures and compliment your fruit, who find a way to navigate perishable peaches into downtown Washington, D.C., are hard to dislike in the flesh and thus so very dangerous to farmers.

Society is a sudden, explosive enemy of agriculture. I have seen forty acres of vineyard down the road, the ancestral home of a fourth-generation family, ripped out, paved over, and planted with over 200 federally subsidized houses in a single summer—all by "a set of men so dangerous." A vineyard becomes an instant suburban nightmare, replete with graffiti, gangs, and low-slung cars sprouting from its south twenty—a guarantee that for the first time in their lives the former farm owners, Mr. and Mrs. Corky Largent, at last have some cash, and that a thousand more of the poorer shower, shave, and watch TV better than 4 billion others on the planet.

I have also seen a beautiful peach orchard uprooted, only to have an exclusive and gated enclave of ersatz Mediterranean homes in its place within the year. Fresno, after all, is the fastest-growing city in one of the fastest-growing states of the fastest-growing decade. This once backwater agrarian center is now a pornopolis renowned for its petty graft, where its white-collar felons boldly advertise their corruption on personalized license plates. Until jailed, one of the more audacious consultants drove his Mercedes with the vanity plates reading REZONE. He was apparently bragging to all about his skill in bribing the local council to bulldoze protected groves of trees and vines—and was sought out throughout his community as a thoroughly good guy who for friends did his consulting without charge. Society can destroy vineyards and orchards, give profit to the destroyer, well-earned retirement to the exhausted yeoman owners—and provide cheap, needed housing for hundreds of impoverished all at once. Who can sort out the morality in all of that? Not I.

Society in its benevolence is a sly enemy of agriculture. I know a man whose now failed empire rose entirely on the failure of others, a man now facing a grand jury for pocketing harvest moneys that were not his. I saw him in town in January. "Do you need crop insurance this fall, Victor?" he asked. I was flattered that he remembered my name, and was about to say, "Now that I think of it, I do." After all, for a few thousand dollars he could protect my crop better than all the thousands of hours of manual labor hired from the working poor for an entire year.

Society is an unstoppable enemy of agriculture. In the last decade over a million family farms were devoured. In the immediate environs of this farm alone, I know of only two vineyards that are larger than twenty acres but smaller than five hundred, where the owner lives from the raisins he grows. The smaller vineyard men go to town, the bigger ones get bigger. The in-betweeners sell out, rent out, or poop out—wives and children happy that the madness has at last ceased and the body can die now in peace. One woman of eighty told me of her husband's demise: "I was happy to sell, just wish we had done it when he was alive and had time to do something with all this money." When the cantankerous weed-fighter finally keels over, the family at least gets that one long-promised cruise. Family farms are usually prized by corporations; and why not? Despite the absence of capital, the previous owners were far better stewards than other corporations.

Society is a shameless enemy of agriculture, lacking the stature of the buck-toothed rabbit, the elemental drive of the scaly hopper, with less beauty than the cocklebur, and less nobility than the filthy nematode. How and why so? Developers with good names like Land Dynamics and United Builders tear peaches out only to call the subsequent lifeless boxes and asphalt Quail Knolls. The uprooted plums of decades become The Orchards. Can't we at least call them, as good communists might, Peoples Subdivision Number 3A, or as postmodernists would, "Nothingness Here," or as empiricists should, "For Sale Now: 47 Houses, two acres of concrete, one acre asphalt, assorted wires, grass, and shrubs"?

Our colleagues are taught to ruin agrarianism at the university only to endow a chair in "agriculture," and then reward the local architects of that onslaught with a massive steel-and-glass "research center." Can we not at least make the nomenclature honest: "The J. G. Gallo Wine Professorship of Corporate Viticulture," given "in appreciation for thousands of days of pro bono corporate consulting that has helped to transform the nature of American agrarianism"? Our good congressman who crafts water and crop subsidies is "agriculture's friend in Washington," a walking, breathing exemption for a few dozen magnates. "Horrible policy," he calls the 1996 farm bill that gives $34 billion to corporate agriculturists like himself. They are the real benefactors of his campaign, he himself the recipient of the very largess he has legislated.

Society takes the necessary but unpleasant from farming and deifies it. The good enemies of agriculture take the need for credit and make the lender Zeus on Olympus, as if a 10 percent interest rate (Mr. Crèvecoeur's "canker-worm which consumes the yearly industry") on a half-million dollars is somehow bad business because our borrower reeks, his nails not altogether fingernails anymore, his half-toothed head not to our liking. The purveyor of chemicals—who is known to pick up trash along the freeway on the regular Rotary Club weekend cleanups—is made a prophet, a salesman now worthy of obeisance because his toxic pouches and bags have racy names and chartreuse, postmodernist designs that earn him $100,000 a year. The exigency of a vast pyramid of trucks, shippers, and peddlers is proof enough to make that overseer with phone and computer into a pharaoh. And those who do such things are our friends at the ballgame, wave to us on the street, and collect and repair bicycles for the underprivileged.

The farmer of this Valley, who alone transforms dirt and water and plants into our food, remains an exploiter of the migrant laborer, the polluter of the wild, or, worse still, the kind yokel with a lined face we know as Gramps on TV. He is the age-old sodbuster who killed the Indian, eroded the prairie, and created the reactionary populism of this nation. Cannot we say instead, "Ray Mix [the seventy-year-old vineyard sprayer] is a tough old hombre, who will do about anything and then some to keep his granddad's ground, whom I sort of admire as long as he keeps himself downwind, this ugly nut who gave all of us so much of what we now take for granted"?

Of all the enemies of agriculture, society, at the millennium, is both its friendliest and most shameless pest. A society that devours its small farmers—and there were many, history teaches us—may deserve, I suppose, what is now happening to us. I learned that not from a page of Hesiod, not even from those odd, now deemed corny, ideas of crazy Messrs. Toynbee and Spengler, surely not at all from any savant within the academy. I learned about the ultimate foe incarnate from all the enemies of agriculture that I now know inhabit or at least visit this tiny farm, and who in their aggregate do not munch, rot, or strangle quite like those outside these borders. The thousands of life-and-death struggles on this farm, human and not; the war right outside this front door to bring fruit from this earth; the ordeal and triumph in wringing one more year from nature—all that taught me more about our adversaries on this planet than all the sociology and psychology I pondered in the university. So with the collapse of agrarianism, I and those who follow will never learn in quite the same way again about whom to fear.

But, thank God, I have also discovered who in this society is not among those infamous enemies of agriculture. Of course, the foe is not the noble owl who preys on the rat and the squirrel in the vineyard, who on ceaseless patrol greets you above the orchard and only occasionally steals away a kitten. Of course, he is not the farm dog, whose teeth, whose very life, are put to service to save you and your kin—and only rarely bites the meter-reader.

But the antagonist of the farmer is also not the darker metic who picks your grapes and fudges on his tray count, who trudges up from Mexico to drive inebriated, to ignore the law, and to work until exhaustion to feed America, only to die as if he had never lived at all. The destroyer of the farm is not your morose son who strips the gears on his first proud day out on the tractor. He is not your brother with whom you argue over the new species of plum to be planted, who fights with you hourly over the loan now called in from the bank. He is not your uncle who asks you to sign his crop loan, pledging that you in fact, or so he thinks, will not lose your house should his peaches rot. He is not your wife who has no time to hear of mad ideas to build a shed, or why you must lose money just to farm. The enemy is not even your pesky farming neighbor who ridicules your short crop of nectarines, and surely he is not the stooped hired man who tells you weekly, "Just you leave me alone today, boss, got it?" He is not, then, Ramon spraying himself as he sprays your vines.

Among the flotsam and jetsam of century-old vines, I have learned that in life, as in agriculture, it is a matter of utmost survival to know, among all those that bother you, who is not the enemy. The badness in this farm, I realized as I stared at the weathered sign of my near-broke neighbor's ranchette, the failing abode of a nice man who talks kindly and yet plowed up our communal road, comes not so much from the bad, however maddening, as it does from the good.

THE GREAT DIVIDE

(For Eaters and Their Providers)

How I hate to dwell in these accumulated and crowded cities!
They are but the confined theatre of cupidity; they exhibit nothing
but the action and reaction of a variety of passions which, being
confined within narrower channels, impel one another with the
greatest vigour. The same passions are more rare in the country;
and, from their greater extent and expansion, they are but neces-
sary gales. I always delight in the country. Have you never felt at
the returning of spring a glow of general pleasure, an indis-
cernible something that pervades our whole frame, an inward in-
voluntary admiration of everything which surrounds us?

—J. Hector St. John de Crèvecoeur,
Sketches of Eighteenth-Century America

I. Country versus City

Most professionals in America always suspect that a farmer and
his ilk are either ignorant or crass, and surely not professionals.
The farmer, in turn, believes that you would have to be crazy to
live in town, crazier still to wear such garb and endure such re-
proach at the office, where the sleek and stupid can as easily excel

as the more real and intelligent fail. Each is the absolute antithesis of the other, the agrarian requiring action above all, the professional anything but force and audacity. Absolutes versus nuance; natural ill manners at odds with studied refinement. The former, of the outdoors, values independence and commitments that are ironclad, the latter, inside, often sees those very ideals as sheer recklessness and obstinacy. The latter likes—no, must have— associations and committees; the former says more than one is too crowded. The rank world of the farmer says, "Don't back down." "Go around," orders the tame cosmos of the urban employee. Is it to be "Let's settle it right now" or "Let's be sure at least to do lunch sometime"? The farmer on his own lives concretely for liberty that can end him, the professional in town for equality that protects him. Country folk fear nature in their midst; city dwellers idolize it at a distance.

No other occupation in America is akin to what we once were than the family farmer, this iconoclast of the countryside, who flees from consensus and equanimity, this oddball who is the stuff of history, with its wars, prejudices, depressions, strifes, and tragedies. The student who visits the farm for a day goes away troubled and unsure whether the world is a logical and kind place, the auditor on campus more than likely hears how steps are being taken to solve the woes of the universe. How odd that Enlightenment thinkers once thought that when America had more teachers than farmers, all the old ignorance would fade, not grow. But grow it did—all the more so.

Clearly, the 99 percent in town do not know in any direct way the 1 percent on the farm; the latter experience the world of the former every day on television, in the mail, and on the phone. And the divide will only get worse, until there is no divide and we are all of the nonfarm world. Agriculture in the millennium to come will for the most part be conducted in vast expanses, away from town—the corporate void where no sane man wishes to live. This vast stretch of latifundia will be as Crèvecoeur's Wild West of grasping, brutal, and antisocial thugs come alive: few Americans in the generation to come will ever see where food is made. Try driving through the millions of uninhabited acres between the 99 and Interstate 5 freeways; that dreary corporate plain of central California and its godless culture nevertheless are responsible for dinner tonight in New York, for the Levi's worn by shoppers in Michigan, and the ketchup at McDonald's in Japan.

In contrast, the agrarian patchwork that surrounds the small town will be absorbed by, or sprinkled with, the suburb, edge city, and ranchette; but, again, real food-making that feeds millions will be safe and far away in factorylike settings. How strange that we will destroy what we love, and save what we despise; how odd— but ultimately sinister—that those who will produce what we eat every day will be mostly invisible, and their workplace unknown. How odd that we will never again know we are urban by not being rural, rural by not being urban. The day is upon us where the eater of a plum, drumstick, or ice cream will have never seen once in his life a fruit tree, chicken, or cow, much less an orchard, henhouse, or dairy. Tell him his steak is made of sawdust and his raisins grown on trees, and he will be as likely to believe it as he will be unperturbed.

When the land matters little, suburbanites and ranchetters will lose the age-old tension between town and countryside, the energy when the rustic came to town and the urbanite drove out to see Gramps. The old clear line between rural and urban was far better than what we are becoming, as we will learn when we are all confused among the sprawl: houses here, empty lots there, shopping centers among a few vineyards, a few orchards somehow in the city limits, organic gardens near city parks—even as the corporate fields of the food business are far away, unmentioned, never visited, and of no aesthetic or cultural interest.

What, then, is so beautiful and thus worth saving in a farm, and how does the agrarian's layout differ from the domain of the magnate? Beauty in farming the countryside comes from symmetry in the aesthetic and human sense, when men organize the wild, which has not yet become the town—the farm becomes the mean between complete artificiality and nature absolute. Tree and vine farming is man's best example of this creation of order from chaos. The planter of trees creates order from nature, in a way even the field-crop grower never can. His duty is to be the border, surrounded by, but separate from, the countryside, the intermediary between the town's concrete and the wild woods and wilderness uninhabitable and unlivable. The farm does not mix the two, but rather separates city and wild. The farmer is civilization's buffer against the untame.

Congruence of orchard, house, alleyway, vineyard, and those who are busy within that cosmos is at the heart of this farm. The borders of trees and vines set off nature and have patterns emulated by floral arrangers, linoleum designers, and wallpaper artists. It turns out that the hard-bitten, sour homesteaders of the past were secretly canvas painters and artistes after all! Their nineteenth-century landscapes did not fade with the harvest, but are still here for the developer to nuance, adopt, and destroy long after they are dead. The most tasteful sojourn in Fresno remains Fig Garden, where private homes in shady lanes were plopped down among a once colossal nineteenth-century fig orchard. No one would call their neighborhood Cotton Acres, Alfalfa Hills, or

Wheat Estates. Nor would a Fig Garden have grown up over in the far distant barracks hamlet of Strathmore, a town snuggled in the West Side of the Valley between the anonymous 50,000-acre land empires of the corporate class.

Our own orchards and vineyard about a hundred yards south of this house are what Cicero meant when he said man can create with his hands another world inside nature's own. Out of bare fields arise order, greenery, culture: squares, rectangles, and triangles of measured arrangement amid natural chaos—human-induced order that tends in turn to have a restraining influence on its creators themselves. Out of open ground you craft a dense canopy of peaches; in response that orchard measures you too, tames you, reinvents the very nature of your own time and space. If you doubt me, walk within a thirty-acre cotton field and feel big and then venture through a like-sized acreage of plums and feel like nothing.

Ten years ago we once planted a plum orchard only to dig out and replant the young trees the next day—the planting wire had proved false in the twilight, and in the clear morning the diagonals of the freshly planted stock appeared out of plumb. Who wants to look for the next thirty years at a slightly crooked row, two trees out of a hundred out of line, a random grove, not an ordained orchard? Children, workers, ourselves, all those who would pick that orchard for the next three decades, might assume from its flaws and imperfections that they too could skip a tree, leave the hard-to-get fruit behind, fail to find order and rhythm from the imperfect consonance of the orchard. Chaos is an infectious agent in agriculture—as it should be in all human endeavors. Farming, then, was the ability to impose order upon nature, to use the thought and labor of mankind to organize nature. The physical outlay of the farm must regiment and thus enhance, not let loose, the power of nature. And even the smallest detail, the lay of the land, the line of a row, matters a great deal.

Near our dry pond is a triangular vineyard that follows the land's contour, 5,000 vines in rows of gradually diminishing length. They are green; the alleyways around are not. The vines are six feet apart within each row, the rows themselves twelve feet from one another. In between is the dirt. There it is over 120 degrees and more in July and August. The ground's dusty color, texture, and temperature are set off by its living verdant antithesis. Nearby are three acres of plums, a rectangle that slopes to the pond, 360 trees in a quincunx pattern—a mosaic of proportioned shapes. "To sell this place, you'd have to pull out all those junky plots, level it, and plant something straight and flat where you weren't turning all day on your tractor," more than one uninvited real estate agent has pontificated. If farming were really but the price of diesel and the wage of the driver, he'd be right. But it is not always—not yet at least. Aristotle, after all, did say that the nicely planned town resembled the pattern of a vineyard, not vice versa.

Likewise, there is contrast among the two acres of pears beside, five acres of pomegranates behind, and also four hundred persimmon trees below. From the air these plots are all jigsaw puzzle pieces of differing hues and shapes. From man's ground view, trunk and limb form diagonals, lines, triangles, and squares in every direction. Is it not the purity of mathematical proportion, Plato's pristine reminder that there is an absolute plan and truth on the other side, discernible only through the divine gift of numbers and proportions, which alone now in our world do not lie?

Trees are planted one after another, no randomness, no missing slot, no three bunched here, one over there alone. Try to imagine an orchard or vineyard not in rows and files. Semicircles? A maze? Squiggley lines? A postmodern hodgepodge? Random groves? It would be impossible to cultivate, difficult to traverse, and a complete refutation of any economical use of space and time. Instead, in their fixed and proportional station, the living grids are constantly rearranged by the eye as parallels, perpendiculars, and tangents. They are man's arithmetic—all on a natural landscape that refreshes and relaxes the soul and mind. I think the latent square, rectangle, and grid are always there; the farmer merely puts his new trees and vines into it. In a way he has no choice.

No wonder the garden of the suburbanite is replete with railroad-tie planters and concrete curbs that set off and accentuate flower and fern from lawn and walkway. Landscapers and gardeners do such things for reasons other than worries over creeping weeds and wasted water. As humans, they too know that the mind abhors chaos, and unwinds only amid order that avoids banal repetition.

But not all the pleasure of the grove's symmetry is mere visual experience. The soul's appreciation of these aesthetic patterns can also arise from the tandem of a working mind and moving body integrated into the natural grid. Drive a tractor among the trees. Tie up vines. Prune an orchard. The labor, at first so very monotonous, turns out not to be drudgery after all—it soon begins to blend you into the grid as well. Plunge into row four, tree eleven, and you learn that you too are part of the quincunx, leaving in your wake borders of pruned peaches next to unpruned, half a vineyard clean, the other half weeded by your disk, rows of vines staked in stark contrast to those left without supports. Instantaneously, the job is one-fifteenth, one-third, or five-eighths finished. By row fifteen you learn that you are more than a vine or tree. You are man, who alone can find and then create such patterns that last but a few hours or an entire lifetime—inside a vineyard or orchard every day, any day of your life on a farm—through your muscle, craft, and tools. We all, do we not, search for some mathematical dimension, some living proportion that is unchanging and so gives us measure as we grow and pass on? Soon this recognition of balance and dimension within the field does not even leave when you do, and thus enhances the cadence of an already rhythmical agricultural calendar. The vineyard's pruning, shredding, tying, disking, suckering, irrigating, harrowing, dusting, picking are timed pulsations of their own. Like the vineyard itself, the work becomes a pattern the farmer learns as he mundanely moves from vine to vine, from row thirteen on through twenty. "Up and back, up and back, mister," old Mohinder Brahr from his truck in the alleyway once barked at me for an hour and more as I was pruning vines. With five hundred acres, a million in the bank—and half a diseased heart for the tab—Mr. Brahr pined still for the stoop labor of others, lost recompense for the unhappiness that the transition from agriculture to agribusiness had brought him. All that money had taken him away from the shears, out of the vineyard, and into the hospital—and now on toxic medicine, slumped over in his new truck. "Up and back, up and back, boy; I pruned all day with twenty men way behind. Up and back. Don't quit now. Come on, prune, up and back—in the dark, like I used to. Always in the dark. Give me the shears, boy, the shears right over here." By the time I made it out of the vineyard with the shears and over to his truck, he was asleep.

Right now I am looking out at the shed. Eight kids—nephews, nieces, son, and daughter—are arranged in a line packing fruit as it leaves the sizer. The Santillan brothers are running the machinery, and Rosalio is cruising through the orchard with loaded bins of peaches. Rigo and his men are on ladders, and Roberto has the truck, checking the irrigation water. Somebody, my brother or cousin maybe, is overseeing this bustle. But like the hoplite general on the right horn of the phalanx, they are indistinguishable from their troops, and at the cutting edge seek no reprieve from heat, dust, and the monotony of the struggle. Everyone doing something different, yet too everyone fit into and ordered by the symmetry of orchard and vineyard that guides the tractor, makes the ladders progress along lines, assures Rigo's pickers that there are no missing trees in the orchard, ensures to Roberto that one irrigation valve is after the next, that the water will go down row 40 like it did 41 and 39 before, that each line in the vineyard according to its station does in fact have 102 vines.

It is quiet out here and it is peaceful in the countryside—but men are at work at a frantic pace to tame, order, and exploit nature. And something is going on—men and women are being reenergized even as they work themselves into exhaustion. Even as they age and wrinkle, the denizens of the farm do so with like kind in the natural world around them, who like them are fighting to nourish themselves one more day and leave something of themselves behind when they leave. Time is both of the essence and irrelevant; we are working for the frantic moment, true, but no differently than the now dead did here five decades ago. No one gets out alive, they say, but we who plant and tame think, in a way, that we do. We like looking at town on our horizon, we like that it is there and not here in our grid.

A farm of trees and vines turns out to be a fairly large, ordered, and understandable universe that exists without conscious worry about others in the city. Its orchards and vineyards say to those wearied by town and disappointed by the failure and rejection that is so much now a part of city life, "Come into me, where you have always belonged. Prune, shovel, or water. Come into me and be me, and I will make you forget all else, forget time itself, forget that you were ever far from me—forget that you were once failed and as nothing." I once went back into these Acres of Forgetfulness in 1980 after nine years pretty much on the outside. I only realized that it was 1985 when an itinerant barn painter offhandedly remarked, as he cleaned out his airless sprayer, "Aren't you the one who went to school over on the coast and was going to be some professor or something?" I snapped to attention from my dream for the first time in sixty months and mused, "You know, I think I was." Watch out for the sirens of the orchard and vineyard.

The balance between symmetry and monotony is tenuous and often lost in town. The exhausted factory worker, despite the reassuring din of choreographed human bustle, can attest to the boredom of the rote and repetition of the assembly line. But when the squares, lines, and rectangles that confront you are alive as well, they engulf more often than wear down. I know what old Rodrigo means now, his mind the clearer, mine the more diseased for previously not grasping his message, when he talks apparent gibberish to my brother. "No worry, we go right down those rows. Hey, outside ones first today for red plums, inside on Wednesday for the rest. Don't worry, we get them all, in the rows with the ladders and buckets." He knows, as I do not, that he is in the grid, in it to stay. I know now what my son meant when I dragged him out of the orchard of Lethe last Saturday to visit a family in San Francisco. He left the quincunx of the Castleman plum orchard, to walk three hours later dazed, lost, and bewildered through the street grids, hammering me with, "What do they do here, how do they live, what's all this for, when can we get home?"

To recuperate from toil and affliction we need some hybrid order in life, some proper proportion between nature let go and the wild silenced, between man full of himself and man without man at all, between beauty that is barren and utility that is ugly—between town and country, order and chaos, work and leisure. We need to see the outline of the city in the distance and know we are not in it. We need to be among the office towers and know that beyond is the farm, that there is a line clearly drawn between us and them— one we all can and must often cross.

We of civilization need to order and grow things that are not ugly and yet not beautifully sterile. We need, each of us, to make things, create things, leave something material behind as products of our arms and backs. We need more than the sensuousness of our bodies and the pleasure of the palate, delights that fade even as we indulge them. We need at one point in our lives to tap that vestigial and ancient gene that comes alive only when we sweat outdoors, our hands at work with and against nature, creating productive beauty that will last beyond ourselves. The city's finely planned intersection, the suburbanites' four ordered rosebushes outside the door, the tasteful hallway that divides the office in two, the tile patterns on the wall of the aesthetic lounge help— but they still do not quite suffice.

Those lines and squares need veins. They should contain water. And they should have a cycle of life and change. Better yet, they should be productive, not merely ornamental. The entire grouping should bend in the wind, invigorate with water, strengthen and sicken by season to remind you what man can do at his best with nature, where he was bred to belong.

Sour tree and vine farmers would deny their sensitivity to order and beauty even as they plant hedgerows, leave in an unprofitable orchard too long, and plant shade trees to crowd and interfere with their sheds and pumps. They say they work in isolation, but know they really do not. They secretly admit that for all their seclusion, they are in the grid that has no time, in a way that the beleaguered office worker amid thousands in town is tragically not. The pruner in the vineyard, like the hawk above veering from invisible quadrant to accustomed quadrant, like the coyote roaming from row to row, feels himself, even if for a moment only, part of this grid, part of others' grid like his own as well. The farm turns out to frame and order and make sense of thousands of exquisite, planned, and interconnected private agendas. So it is curative. Take the worst office day in your life—losing a job, savaged by boss and overseer, ridiculed by administrative peer and rival alike, forlorn in love—the wounds can all heal in minutes, if you wade into a grid of trees and prune until you are with them. Twenty years ago a man at the university with various degrees and some good cause told me I could not think. I worried and sulked about it for a week in my East Palo Alto apartment—until I drove back to this farm, waded into a vineyard outside the window where I now write, and instantaneously forgot that he was alive—forever.

Critics of family farmers are right that we are not as efficient as agribusiness in providing great quantities of food instantaneously from great distances—we lack the capital, the trucks, stores, brokers, salesmen, that bring safe, tasteless, and affordable fruit at all seasons from anywhere. We do crazy things like charge no interest for loans to neighbors and friends. We live in a world of our own making that we call moral, what others call unrealistic if not uneconomical. But if in the past the rural world of our grandfathers' society was too moral and therefore uneconomical, perhaps now in the urban cosmos of their children we are too efficient due to our very amorality. So if a hundred years ago farmers and their nonsensical world of toil, shame, and isolation were a drag on the progress to a cure for brain cancer, penal reform, and watching the World Series on blurry television, perhaps now the collapse of agrarianism helps explain the acceptance of liposuction, murderers on parole, and call-in sex talk live from New York.

Not long ago four gang-bangers shot and killed a farmer down the road in this Valley. Four miles away from this house, they broke in and pistol-whipped his wife and robbed them both. The murdered victim—he, of course, chose to resist—farmed for over sixty years, only to be slain in his ninth decade for a few bills in his wallet and the ring on his wife's finger. I think it not too cruel to say that he was a better man than his assailants, his family better than theirs, his grid noble, theirs savage and cruel. And I think it is not too utopian to say that had they grown up in his agrarian grid, they would be pruning, not shooting a pruner that day. You see, he was the best of what the country might produce, they the worst the town has destroyed.

Without an orchard and its ordered lines of limbs and trunks, without vineyard grids to wade into each evening, to make you forget time and space of the city, how could anyone—how can you right now, reader—parry such chaos? Where will be the one to assure you that there was once a different way, that you, city man, in your hour of present discontent, in the prime of your terribly efficient economy, are not crazy?

II. Utility or Sterility?

We pave our alleyways with culled fruit. It was my idea years ago—after seeing it done elsewhere—to use the plum, peach, and apricot rejects to keep down the dust. We needed to find a place to throw the ruin, and thereby to save on road oil and dumping fees.

My son once thought he saw a bird eating the rot, so there too the detritus might save a peck or two from the orchard, as crows ate garbage on the ground rather than food in the tree.

Those fruity alleyways are at least testament to the power of the cultivated trees on either side, whose natural exuberance, once tamed and bridled by us, direct their energies to the most amazing purposes. You see, each year 30 percent of all the plums, peaches, pears, nectarines, and apricots on this farm, and in the United States, have blemishes. These mars range from the serious—wormholes, brown rot, bird pecks, underripeness, and fingernail wounds from the pickers—to the insignificant and trivial: sheep-nosed rather than round, wrong color, wrong texture, slightly overripe, or wind-scarred. The latter category of imperfection affects the taste and quality not at all. But the broker is right: tons of unsightly, misshapen nutritious fruit sit unpurchased by the discriminating shopper, who must look for appearance rather than taste.

So we throw culled fruit on the dirt alleyways to keep down the dust and save the hauling fee. If any of you want it, please stop by. Rent a U-Haul, pack up a ton or two, and haul off enough delicious off-grade nectarines for your entire town before the fruit goes bad in a day. There is enough jam and preserves for an entire city each day on our roads. The trees on this farm produce daily tons of nutritious unwanted produce that was grown, watered, fertilized, picked, transported, sized, culled—and dumped. In other words, America's fruit farms feed the nation and world, and throw away enough to feed a rather sizable planet to boot.

The tons that head east from here in boxes—and the tons that never do, but rot instead on the ground beneath and beside where they were grown—give some idea of the power of the farmer's cultivated plants and the potential of a cultured countryside. Peaches can produce thirty tons and more per acre, a mere 120 trees filling up one semi-truck and a half with fruit—an acre of peaches providing 50,000 city folk a fat ripe peach each day. Thompson vines sometimes top out at fifteen tons green per acre or three tons dry. One farm of sixty acres can furnish all of Fresno its per capita raisin consumption for a year. Each acre makes possible dozens of lives in the city; the countryside is in fact a very utilitarian nurse to millions who will never see it.

It is magical, this natural factory of plant, water, air, and soil that pumps out beautiful fruit by the ton. Mr. Crèvecoeur's empowered farmer knows this and can be ruthless in his single-minded attempt to accelerate that natural process, to squeeze everything from the beautiful countryside. He prunes enough wood to ensure well-sized fruit, but not too much to bring on scarcity. He waters frequently, but not to an excess that can split swollen and ripening peaches and spur unwanted, useless vegetative growth on his vines. Manure, sometimes at ten ton an acre and more, is spread over and disked in the soil—not enough to burn leaves, yet adequate to turn all the plant's reserves to fruit augmentation. When the farmer intervenes to ensure that the orchard and vineyard have no enemy on leaf or root, no sucker, biter, or chomper that destroys chlorophyll, then the plant is at last turned loose to produce fruit.

"Take a look at old Perfect Farmer down the road," my neighbors told me of the local raisin king, Charlie Arakeian. "His wires are breaking under the load and it's not even August first." Had the passerby contemplated in depth Mr. Arakeian's grape bounty, he would have at last surmised what the poor fellow was doing all winter and spring—fertilizing, watering, and pruning according to intricate mystic hypothesis and private conjecture. Charlie scarcely can walk these days, but the old maestro hobbles out to his vineyard to tell his hired man where and how to pour on the water and nitrogen, for it's the grape he's after, far more bunches than you, I, or he will ever need. It's the power of the vine that interests him, the beauty that accrues when sweet grapes sprout from gnarly stumps and fifty-year-old stock. He likes three tons of raisins drying per acre, 1,500 trays of drying grapes that scarcely fit between his rows on every acre he owns. You must understand, like some jet test pilot or theoretical physicist, he is after the unattainable, in search of the hypothetical maximum, the quest to learn just how many quality grapes a vineyard might produce if every factor was optimum, if the farmer was God. Charlie finds beauty in making those ancestral vines work like none other. He lives for the bunches that sag the wire.

The results at fruit harvest on a farm are sometimes staggering! Nectarine and peach trees are roped to prevent limbs from snapping under the weight. And still the fruit weighs down, causing the farmer to bring out wooden tree props to brace the limbs until harvest. Occasionally even that is not enough, as whole trees simply topple over, 800 and 1,000 pounds of produce that not even the massive roots, ropes, or props can support. Did nature ever intend for such trees to put forth such fruit? Hardly.

The orchard then almost—though not quite, since most trees in the orchard do not uproot—reaches deformity, as the trees' sheer power approaches the limitations of their genetic imprint. The farmer, not the chemist or biologist, coaxes forth the bounty through skill and labor, not test tubes and microscopes. The orchard's power and training stop just short of that of the Olympic bodybuilder, whose brawn and ripples finally incur repulsion as he hobbles under the weight of his own muscular mass, or the monkeylike thirteen-year-old gymnast, whose stark regimen has arrested puberty itself and embraced androgyny as the price of twirling around a steel bar. So too the roped and propped peach tree occasionally pops out of the ground and is killed by its own cultivation and success.

The power of fruit trees and vines is regimented by the farmer. Should he prune too much, thin excessively, fertilize redundantly—or lose his blossoms to frost or hail—then the crop that is left is too small to manage the power of the vine or tree. Leaves predominate along with shoots and tendrils. What fruit there is often cracks, and is even too big, too mushy and watery—tasteless even.

But do not prune, do not thin, do not fertilize, either out of ignorance, a sense of economy, sloth—or greed—and the plant rebels. It reverts back to its natural imprint, which is to produce thousands of fruit, not to be eaten per se, but through untold pits to reproduce its own genetic code. Thousands of sour, undersized, and tasteless berries wear down the poor tree and vine, unmarketable, inedible, unsightly, and with far too much pulp for that poor overtaxed tree and vine to sweeten and size.

Trees and vines are not like our state's stunning giant Sequoias, the bristlecone pine, or sacrosanct barrel cactus, California's rare and majestic natural flora, whose beauty, age, and scarcity earn divine status but otherwise do little to feed, clothe, or shelter the citizenry. Cultivated fruits are beautiful through the sheer powers of production, of the unending natural sugar that appears from such lowly wood and leaves. The power to save people through man's intervention in nature turns out to be aesthetic in its own right: one tree—five hundred fresh-fruit eaters for a week; one vineyard—a small town in Ohio's yearly supply of raisins; an orchard of persimmons—all of San Francisco Chinatown's winter supply of fresh fruit.

The productive and healthy vineyard and orchard entice you at harvest, as you relearn through their fruits that they are not the flowering pear of the apartment complex. They are not the fragile

Japanese maples that line the dentist's office, much less the blooming but toxic oleander along the highway. Trees and vines at picking time make all sterile plants in contrast appear as nothing but decorative ornaments. In a California full of majestic coast redwoods without the nutritional power of the Valley's orchards and vineyards, we would all likely starve in Paradise.

Our farms are beautiful still because they are for a while longer powerfully useful. Tourists in ignorance pass them by on their way up to Yosemite. They speed past the floral bloom of millions of plums soon to be, only to see trees and rocks that feed no one. The countryside is not the wilderness, but man's natural checkerboard that quietly, each day, with each irrigation, gave life unlimited to millions of urban others.

The battle is not just between farmer and farm, but pits the farmer against the produce broker as the former struggles against his instincts to find the mean between quality and quantity. It changes yearly—no, daily—depending on the weather, the stock, the soil, the market, and the farmer. My grandfather once summed up farming as the quest to put out the largest possible crop a tree or vine can still bring to successful maturity—a task no living mortal can accomplish.

The too cautious farmer, overly worried about quality, never quite achieves the production necessary to save his ranch. He brags about compliments on his sterling produce and he does crow about the price per box he receives. But this blinkered perfectionist grows silent when pressed to enumerate exactly how many of such matchless boxes per acre he grows. Yes, the brokers love him and his matchless product. The banks praise his devotion to excellence as they nevertheless pore over his pathetic production readouts to reject his loan application for next year's crop.

In contrast, the far more sly, reckless, and rapacious grower wants only the tonnage, not quality. His pruners leave too much fruitwood; he does not thin his plums; he does not worry that his tiny peaches are but an inch apart on the limb. He even runs out of tree props to support the cherrylike plums on thousands of limbs. His legacy? Tons of tiny sour junk that no one wants. And this grasping fool lectures the gullible media about all the food (that he needlessly created) on his farm that no one eats: "Why, they, I mean the plum board, won't let me sell my own darn harvests. Even if it is small and a little sour, the plum's plenty good to eat, and I got enough for everyone to have all they want." Then the media interview ends, and the producer of garbage resumes his doomed quest to convince the local supermarket to allow him to back his semi-truck in, to unload the unpacked refuse from the field, and at last to allow the consumer to munch all he wants of unattractive, bitter, small, and rather ugly peaches and plums. This sloppy, duplicitous farmer, in his last and worst manifestation, poses as sole friend to the consumer!

III. Ancient and Modern

In a morning on a farm you check the oil in the Massey Ferguson. Hitch the tractor's disk. Plug the irrigation pipe leak. Fix ladders. Sit in the orchard for lunch. Blow your nose while shoveling. Wade through the orchard underwater. Cultivate, repair, prune, and water constantly. Time is the rub. Machines must be fueled, greased, and driven to meet nature right now. Coddling moths are in the air. If not sprayed, in twenty-four hours they will stealthily inject their eggs into your pears, the hatched worm at harvest eating his way, unseen, from the core to the surface. Months later, transformed from moth egg to worm, his ugly head breaks the fruit's dermas just when your loyal shopper says, "Boy, these Asian pears are good, and pretty too."

There are times in farming when hours—no, minutes—count. All the while in that frenzy, the only eyes you meet are not human and care little for your look. All goes well before the afternoon heat arrives—as long as under any circumstances you do not bolt into town for a replacement spray nozzle or new battery. Do that, bumpkin, and a keen-eyed, sour chorus of heckling nitpickers awaits.

Most off the farm—and that is about everybody nowadays—will stare and slink away from you—you the farmer, who in his wild morning regimen has forgotten how he appears to the vast majority in town. Does not farming, they have heard, have at least something better than this—some agribusiness graduates with penholders, pagers, and moussed hair?

Those in line at the bank, on the sidewalk, at the store, suddenly fathom that you, a grown man, have just hours ago wiped the dipstick on your pant's leg, or have this morning's bearing grease on your hands and wrists. Is there a whiff of pesticide in the air, a veritable mobile, breathing aerosol can of Raid walking about? Or is it this reeking farmer now in our vicinity? From the knee down you are muddy and slushy. Do you have dirt on your backside, mucus on your cuff, and in general apparently know neither comb nor hat? And the braver whose eyes follow you out to the truck are repelled further by the detritus in your pickup's bed: pipe, nails, rusted valves, twine, cans, and dirt everywhere as it should be—a mobile life raft of accessories and parts that appear as an itinerant trash heap to the uninitiated. I, a "Professor of Classics," once parked my pickup next to a professor of farming at the local university. I think he was repelled by me and my worn Dodge, as I know I was by him—a real *doctus* of agricultural theory, whose truck had a bra, bedliner, superfluous lights, and roll bar, but no apparent agricultural use or purpose I could fathom. A cardboard happy face was unfolded over the windshield to protect the dash, and no dust, much less dirt, spotted his hood. All four hubcaps, even, were in place!

The antithesis between us and them, the very few in the countryside and most in the city, goes on for just a bit longer still. People must realize that farming is not statistics from the Department of Agriculture nor quotes from some scrubbed agribusiness talking head, but rather a very unpleasant and brutal task to bring food out of the dirt. Farmers are not studied grungers, not avant-garde trashy-looking students, not connoisseurs of sophisticated catalogues of trendy outdoor apparel. People must realize that their unkempt hair, torn clothes, and a rather rough mien are not credentials to prove sophisticated nihilism or to reflect a vaguely leftist critique of suburban polyester or to exhibit (cheaply and at a distance) allegiance with the poorer and exploited. We do not feel guilty about what we do, and we make no apologies for ourselves and those like us—our appearance at least suggests that. Sometimes rather dirty and ugly men are the custodians of verdant orchards and idyllic vineyards. To keep either growing, weed-free, and irrigated, there is a measure of brutality that even the most humane farmer cannot escape nor hide.

This divide between countryside and town is more than mere appearance and also involves the binaries of itinerancy and roots, volatility and permanence, tribalism and the *polis*. Could the following fabricated obituary for a man of the city ever be true now? "Mr. Rex Smith, an executive for an advertising agency, died recently at seventy-eight. He lived his entire life in the house where he was born, and worked for fifty years at the local firm, where his son and daughter, who live on either side of the original Smith home, now are also employed, as well as both of his granddaughters."

Or try to envision the reverse: "Royce Smethers, a farmer, passed away recently at seventy-eight. He was transferred to the Valley from Utah and retired twenty years ago after a successful career in vine service in Fresno, plum management in Selma, and nectarine oversight in Reedley, where he was promoted to executive vice president for field operations. His career started in Bakersfield as assistant for market advancement for melons, before joining Tree World's apricot operations team in Merced. For a decade he was vice president of sales for Agrisun in New York. He leaves behind children in Texas, New York, and Utah."

We expect all others in America to be itinerant and to be in search of cash, always to be buried where they were not born, to live apart from their children, to gauge their worth in the manner of a *cursus honorum* of some lupine Roman aedile. In contrast, the farmer we assume to be a food producer, true. But he is also a man whose birth, death, residence, and progeny are fated at the moment of his conception—if he, in fact, is to be a farmer. And his rootedness stands as a rock amid the tide of others, as they splash by him to and fro, wearing him, eroding him, but always, inevitably retreating before him, to leave him there and alone but a second after they had by all appearances washed him away.

Western culture began with the rise of this ideology of agrarianism. It was the notion that communities of small farmers would craft their own laws, fight their own wars, and own land on which to do as they pleased, inventing the concept of a citizen and freedom itself. No other culture had free citizens before the city-state emerged—the very word "citizen" does not exist in the non-

Greek vocabulary of the Mediterranean. Few have had them since. But within two centuries of the discovery of the *polis,* the ancient Greeks were facing the contradictions of their own success. They were now free, increasingly prosperous, often educated—and not hungry. Agricultural surpluses and the stability of the countryside under this revolutionary decentralized and private regime thus gave way to trade, commerce, urbanism, and travel—and then the beneficiaries in turn created leisure, affluence, and liberality, as rural egalitarianism became radical urban democracy, freedom smugness and at times cynicism, independence self-centeredness, and prosperity overindulgence—such are the wages when men are free to think, work, and prosper, by law safe from thug, pharaoh, or prophet. Yet this second and subsequent Western tradition of unmatched material accomplishment—the legacy of the Hellenistic, not the Hellenic, world—saw all moral impediments to economic practice, to parochial and quasi-ritualized infantry warfare, to desire itself, let loose. The natural classical genius that was found in materialism and individualism was left unchecked, and so Western man was on his way to becoming Hellenistic and imperial—on his way to seeing that we were, above all, fat, safe, and complacent.

In other words, the city-state—the city and its surrounding cultivated land—was now to be more the city than the state, and what had once been seen as salvation was now deemed passé, if not harsh and contradictory. The Greeks of the later fifth century and the fourth century did not need the aftermath of World War I to discover something of the nihilism of modernism. There were sophists long before Michel Foucault; ask Aristophanes whether there were enough affluent, nasal-voiced, long-haired elite dandies mad at their parents, their gods, and their culture well before comparative literature graduate seminars at Stanford.

So the rise of an urban culture, as dynamic as it was explosive, was often antithetical to, rather than complementary of, early rural values, themselves rather absolute and blinkered—and, to be frank, confining. Within two centuries of the arrival of the Greek city-state, a paradox appeared that plagues us still in the West: the more stable, the more prosperous, the more law-abiding, and the more humane becomes the society that our founders built with their ideas of individual freedom and constitutional government, the further their offspring move away from the sacrifice and hardship that were responsible for that initial and thereafter necessary bounty—a natural paradox of aging, as the romantic Roman poets like Lucretius, Horace, and Virgil knew, that is best seen in the move from rural to urban. We farmers of the countryside create the pleasant circumstances that can alone bring on our own self-loathing. Not a few historians thought from creation to decline took but three generations. Juvenal, Tacitus, and Petronius, the great skeptical triad of Roman imperial genius, knew of the smugness that arose out of the embarrassment of riches.

Philosophy, oratory, tragedy, high art, and comedy are the dividends of a sound social fabric. But they are also inevitably the products of an urban and elite culture, which sometimes despises the rather unsophisticated arms and backs that make it all possible in the first place. Solon's farmers of Attica would never have given us a gaudy Acropolis, but then they were not the urban mob who voted to send the classical Athenian armada to its ruin at Sicily either, or who slaughtered autonomous islanders to pay for the Parthenon or nodded their heads while they voted to execute Socrates. In any case, it was the yokels outside the city wall and their unremembered creation of free speech, constitutional practice, and economic activity in the first place that made the later more democratic city-state capable of both artistic creation and ruthless destruction. There is no solution to this tragic, endless cycle of creation, enjoyment, and decline in the West. Would that we could arrest the evolution in mid-cycle, and hold on to a well-read agrarian or a callused philosopher. Would that we might retain picture-book city-states, nestled in small valleys, the agrarian patchwork of homestead farms spread about, everyone content to have town and countryside separate but in harmony, none eager to have the two mixed but in cacophony.

Dutiful men and women who unceasingly labor with like kind on the farm, marry, raise children, craft and obey laws, apparently do so in order that others to come may choose not to—and that others may often resent those who allow them that very choice. In the West they create the scaffolds that support the painters and writers above. For the ancients—Hesiod, Virgil, and others were onto it—the cycle of toil, leisure, and degeneration was biological, the state an organism with a very definite life cycle. Culture is created in the countryside through hard work and sacrifice, which leads to a better life in town, where the rough and often cruel edges are sanded, where the resulting polish softens our progeny—often to such a degree that original victory over the wild is now meaningless or forgotten entirely, or finally ridiculed as quaint and silly by a cynical, smug, and immobilized third generation—or worse, still romanticized as a day in the garden at a villa on Capri. The bedraggled yeomen at Cannae were slaughtered so that men like Petronius' Giton and Encolpius one day could freely and without worry prance at the baths with all sexes or practice their farting at elegant dinner parties—or laugh at the industrious rag-collectors and strong-jawed centurions still in their midst.

Jefferson's America was 85 percent agrarian, and he alone seems to have predicted that such a society could not exist as it once had when 95 percent lived in the city. How could it when the original idea was to demand of American citizens that they bring to government their distrust of complexity and bureaucracy, their reliance on self and family, their faith in their own arms and head, their knowledge of the fragility of nature harnessed and the chaos of a nature let go, their skepticism of taxation and of the idea that someone who does not grow, manufacture, or build could ever know best how money should be spent? How could it be that Crèvecoeur's man, who had muscles and courage and an innate aesthetic sense of and respect for the wild, would prefer to see those traits become vestigial in the city? Are we surprised that 6 million crowded into Hong Kong, Tokyo, or Singapore are more or less well-behaved, a similar number spread out in Los Angeles—still imbued with the tradition begun by Mr. Crèvecoeur's independent planter yeoman—bored, restless, and dangerous? Ignorant Mr. Crèvecoeur back from the dead would need no more than a glance at L.A. to pronounce a failure of the human spirit— and would he be wrong?

By the fifth century B.C. there was a clear antithesis at Athens between the rustic (*agroikos*) and the more urbane (*asteios*). It was a stock theme in Aristophanes' comedies, where naive buffoons frequently outsmart the more sophisticated who had lost touch with natural pragmatism, who were smart but not wise. Aristotle and Plato, who sought an elusive middle between abstraction and pragmatism, must have felt the divide keenly. Although both men were products of the *polis* and the intellectual electricity of urban living, they were nevertheless at heart conservatives. Both may have been maladjusted and reactionary. Both dreamed of a pre-Athenian world, where men still fought as hoplites, worked the soil, and saw the world in absolutes, immune from distortions in language and the loss of shame. So in both their utopias, land-holding is central. Plato and Aristotle even offer the unworkable solution (as my parents learned) that the ideal citizen is to have two homes—one in the countryside, one in the city—as if intellectual contemplation could be tempered and gauged against rural savvy, as if urban sophistication could be grounded with an occasional infusion of rural pragmatism.

This polarity between city and countryside, profit and sustenance, leisure and drudgery, the stuff of all Western civilization, has been but a smaller skirmish in the wider question, "What is wisdom?" Is knowledge—not the accumulation of facts—to be the accrued body of erudition from the ages, abstract and printed, the academic sweep from technology to aesthetics? Or is it found alone in the school of hard knocks, the experience drawn from mechanics and fabrication with the hands, the wisdom gained from thousands of personal misfortunes in the social and natural jungle, uncontaminated by pampered abstraction? Does a man understand the universe because he can read Descartes, or does such insight arise only after he has lost his ripe crop a day before harvest? Is the university to be 1,000 peach trees, black soil, and a sharp March wind from the north, or the dreary seminar room in the Lit. Department? Is the scholar with hands polished by the page the true visionary, or is it old Ray Mix, with his missing limb and the remaining hand just one large callous for all the extra duty? Do you send young Jason or Nicole to yet another year of computer camp and SAT preparatory study, or put them to work in the orchard for summer?

"Both," Socrates, killer at the battle of Delium, inquisitor of Protagoras, would say. To the classical Greeks, for whom our modern notion of an institutionalized academia did not exist, abstraction was to be married within the same person with action, contemplation, and theory immediately substantiated or rejected through praxis. Too much reading created useless and unread data and knowledge of no value that would eventually lead to the creation of scholastic, academic, and pedantic expression of no interest or purpose to anyone. After all, I could write a lengthy and unread history of every telephone pole on this farm and have the monograph blurbed by obsequious peers as "cutting-edge," "visionary," and "much-needed." More blurbs and reviews still if I included the words "rhetoric," "gender," or "construct" in the thesis title. More so even yet, should I footnote all the blurbers and reviewers inside.

A $30,000 fellowship can allow a young scholar the chance to provide America with a fuller appreciation of a fragmented epigram of the Hellenistic poet Callimachus. But that gift does not mean that either the nation's treasury or the young man's time was well spent, or that the scholar or his country was more knowledgeable or humane for the investment. The appreciation of ambiguity and irony that education spawns, much less the reservoir of accumulated data that accrues, is not always an aid to the moral life. Just as likely the wages of erudition are a lost sense of morality: when we learn too much, we can talk away, think away, and rationalize away any vice we like. I am afraid that I now agree with Mr. Crèvecoeur about the ideal of a nation of "scholars of husbandry":

> As I intend my children neither for the law nor the church, but for the cultivation of the land, I wish them no literary accomplishments; I pray heaven that they may be one day nothing more than expert scholars in husbandry; this is the

science which made our continent to flourish more rapidly than any other.

Oratory, logic, and abstraction are but a few of the tools the intellectual acquires to make the down up, the good bad. *Sophos,* wise man, is but a hairsbreadth from *sophistikos,* wise guy, checked and kept on the right side only by muscles and toil and unthinking drudgery on occasion. Professors, artists, pundits, you really do need plumbers and carpenters—and farmers—around to tell you that you are often full of crap, that you should not explain away what you can explain away. The muscular classes are more valuable to your work than all the money in Mr. Guggenheim's will.

True, with too much labor, the exhausted mind could not make sense or purpose of the daily wear and damage to the flesh. If you prune, water, and cultivate trees and vines all day for a year, sometimes an hour with a book can cause migraines. No wonder Aristotle labored over the proper mixture of the Man of Action with the Man of Contemplation, and felt farmers had too much to do to become truly politic. Only for a while did the Greeks join the pragmatic with the sophisticated, and only with difficulty did the bumpkin come into town to hear Demosthenes or clap for Sophocles. Believe me, it is hard to do that, and so it is rare to have true family farms and the world of agrarianism around for very long, these strange men and women who themselves must think how to employ their own muscles for their own survival, these strange people who seemingly pop out and rise from the soil we ignore.

Yet that active mind tempered by muscles creates a tension at the heart of knowledge. And it is a balance almost impossible to maintain now, since city and countryside are more often at war rather than at peace with each other—or, worse, since there is increasingly little countryside left. We are now Hittites and Egyptians, perhaps Aztecs too, rarely any longer Argives, Eleans, and Thebans of a teeming rural patchwork.

We country dwellers, you see, can go to town, can see what we need and learn what we lack. But is the opposite any longer true? Can the city boy harrow for a few hours each week, shovel each summer, get peach fuzz down his back as tonic to his urban existence? We know how our peaches are sold, the color of our raisin boxes, how much work it takes to drive plums to New York and cart them off to Food Mart, how to turn on the Internet to learn the hourly price of pomegranates.

But do you, reader, understand the feel of shears in your hands for a month? The trick of picking from a three-legged ladder with fifty pounds of apricots strapped to your belly? Of pulling a tandem disk for ten hours, six inches away from the vines? The frustration of trying to figure out why your five long and tiring days of fertilizing destroyed your plum crop instead of saving it? The embarrassment of explaining to Mr. Ulysses Ponce and his crew of fifteen at 6 A.M. in the morning that, in fact, you have erred, the plums are not quite ripe, and can he and his rather impoverished pickers simply go away for the day unpaid and come back tomorrow, same time? The knowledge that you are doing so to a Mr. Ponce, whose own hands are callused and who pledged faith to those who took him at his word? Do you know how to fail on your own? Do you know that for all your intelligence a weak arm can at this late age bring ruin still? Or that for all those pectorals a simple error of mental reckoning loses your pomegranates? That a day inside with the flu means a day that vines were not watered? Or that a vile, mean man cares not a whit that you are right, rational, and sincere, if he senses you are weak? Vines and orchards, then, are not the 1,000 trees or 20,000 vines to the naked eye, but thousands of mundane tasks and divine revelations that inculcate generation after generation of wisdom that has no calibration, real learning that has no doctorate or master of anything.

Should pruners leave four vine canes or six? Or perhaps even eight? Four should result in good-quality grapes in the fall, but less of a crop—and in some years no crop at all. Six mean if it should freeze or hail after pruning, some grapes will still be there at harvest. When there are eight canes to wrap on the wires, you ensure that even sloppy vine tiers cannot damage your crop by breaking off canes. But then a good spring, free of disaster, just may result in seventy bunches in September, not twenty, making you the greedy fool with a large, sour, and worthless vintage on your now embarrassingly bushy eight canes. In March, should you rise at 3 A.M. to pump expensive warm subterranean water down your vine rows, to guarantee against crop-destroying frost? Some sleepless and exhausted farmers do just that and lose their crop anyway, while their snoring and tightfisted neighbor's vineyard, weedy and unready for the cold nights, comes through unscathed every season—the divine, not man, has decided his place and not yours is by nature always a degree or two warmer than freezing. In your sixties and without a pension, do you "rent" your vineyard *gratis* to your sober, hardworking, and broke son-in-law, who needs 100, not 70, percent of the crop from your ground to feed your grandchildren? And should you rent forty, sixty, or a hundred beehives for your almond grove, knowing that forty might be enough to pollinate a crop, that the safe path of a hundred is well beyond your budget, that sixty gives a good chance of success—and in a warm year perhaps too much success and thus too many nuts. Or is the proper solution—and there is always a proper solution—somewhere between forty-two and fifty-one hives after all? Do you still wish to buy exorbitant hail insurance for your plums from the slick broker of dubious character, whom you have enriched each year with cash premiums on the slight chance that once or not even once in this life a March rain will turn hard and scar and knock off your fruit? The bank said that you should protect its money, but then lends you no money to do so; the weatherman says hail is unlikely, but then will not be foreclosed upon when he is wrong; your brother, you suspect, thinks you gullible for doing so, and just maybe idiotic for not doing so; your daughter needs that premium for tuition money now, but won't be in college at all if it hails. And so you always or never buy the Guardian Plum Plan? Or buy it for a decade when the skies are clear, skipping it the one spring the black clouds shred your orchards? And when your vines are scorched and the irrigation district's water is short, out of principle and in fear of divine wrath, do you take exactly and only your rightful turn on the communal ditch? Such rectitude means that you will cover 60 percent of your vineyard, so that your pesky neighbor down the line, who is prone to take a little too much of a little too long a turn, will now have enough hydraulic pressure to hog the ditch and thereby save his grapes and not yours?

Can we Americans, then, as we used to, and as the Greeks taught us, any longer mold the complete citizen, who—like Pericles, Socrates, or Sophocles—could wound, sail, build, plant, and chisel between speeches, plays, and debates? In short, we need town and city, which are nothing without each other. We have the latter of sorts, the increasingly specialized and narrow, to surfeit. But as for the former, is it tapped for its knowledge, is an active rural life even there any longer to be had? Are there working citizens outside the city limits? Are there city limits at all? The Greeks, who unlike us were seldom obese and occasionally even were hungry, knew that man farms not merely to be fed, but also to learn how his society should be organized. We now farm to eat cheaply (as if America's ongoing problem is famine or an absence of disposable income, as if Americans are too thin), and so have lost the best—and is it the only?—blueprint of how we are to organize a society.

The farm and the countryside even now say to their offspring: "You shall learn independence, for on you alone the trees and vines depend for their culture and nourishment. Your machines, your men, your buildings, your plants are not for others to tend but look to you alone. Fail, and who but you deserves the dividends of that deficiency? Like your trees and vines, you learn man is a natural animal who grows and dies—but unlike all nature, he must use muscles and physical power in tandem with intellect to alter his environment. Your hired man, your father, you yourself sometimes will act, act wrongly until stopped, often at last resort only through the force of muscle itself. The water you drink, the waste that you excrete, shall come out of and go back to your ground. Fail to pump out your septic tank? Do not service your pump? Your family itself grinds to a halt. You may know exactly how your tractor creates power, how the transmission needs professional care, but if your so creative and skilled mind cannot endure ten hours and more of unthinking, loud, hot, and boring driving on such a machine, your knowledge is nothing. You may hate your neighbor, find your environs dull and oppressive, but do not forget that you are here because others of about the same station, of about the same class, are here as well as you.

"You were born and shall die—be buried too—in the same soil. You have kindred that are like you, who once slept in your very bed and whose corpses rot in your ground, and to whom you owe and from whom in turn can demand. And so you need few others than your own, and you will have no comfort around those who are not as you. You, who can take care of yourself, are thus conservative, and intolerant, and expect all others to be, or at least to try to be, as yourself. Bring that wisdom such as it is to town and then leave. Do not live among people, but do not live completely without them either."

The infection of the urban virus in you replies. "Parochial, timid, and small-minded fool! There are fountains, museums, and temples in this world that are somewhat more beautiful than a sunset over the orchard. What is the knowledge of grafting, pruning, even the synchronization of body into the cycle of nature without abstraction, which alone makes sense of such rote? You may learn of patience as you tie five hundred vines an acre up to the stake, but that same lesson of dutifulness, with the added delights of language, meter, cadence, imagery as well, can be imbued through Plato or Shakespeare or Keats. The snotty sleeve is but the crust of the savage. The true trick is to keep the embroidered cuff clean in a world where most things are scum. If civilization were you and your farm, this countryside would be but one enormous Amish compound, a picture-book, crime-free world that ended at dusk. Try curing your wife's tumor without going into town. In your world the Greeks would be but Thebans, their legacy no Socrates, no Thucydides, no democracy, but at best a blunt, nutty Hesiod, a few rustic temples of limestone with stucco veneer, and government by property owners alone. Even your hero Epaminondas was no Pericles; even old man Cato was no Cicero.

"No, it is the city for which we are to exist, our work, our food, the city where change and innovation spark, the city which is for a higher goal beyond simple acquisition and consumption, the city which takes the fuel of the countryside and spits out art, literature, science, and culture. The country's food is ultimately for the urban mind, whose arts, whose inventions, whose music takes us away from the brutality of the rural daily drudge. We do not work with our muscles to eat, but to eat and then think about not using our muscles. No wonder silly Mr. Braudel called the city, not the countryside, 'the transformers' where the electricity of life is enhanced and passed on, not shorted out among voiceless trees and vines, and provincial rustics.'

The farm, I think, is Sophoclean, the city Euripidean. The former is an acknowledgment that our absolute code of good and evil, our adherence to an outdated code, has no place within the present world. We are in the countryside at the millennium now to be tragic, to pass up the better money in town, to live in an old house when it would be better to tear it down and move on, to farm until we go broke, to fashion and adhere to a very old regimen that is teaching us as it kills us, to shovel and prune in silence as we measure out eighty years by vine rows and irrigation valves turned on three turns and then off again three.

In the present age we all eventually come to rue our stance, which gives us permanence, roots, and a creed that most others now will not be as we—but no happiness, no success in the present material world. We choose to be cut off from the national ambition of what America is to be now: mobile, affluent, leisured, and absolutely without obligations—free from shame, free from commitment, free from everything that makes life hard and often boring as well. And so we agrarian absolutists are immature, stubborn, and hard to get along with, know-it-all Antigones and Ajaxes who want to do it the old way—our way—or no way at all. Damn the conse-

quences. Too bad for those we must support and care for. We are
not courageous—there can be real mettle involved in changing and
moving for the interests of others—but we are at least unchanging,
and in today's world that, I suppose, is something.

We on the farm, in other words, have been left behind and
do not fit and do not reproduce ourselves. I have a feeling that
even the aged Sophocles did not like the direction Athens was
headed, did not really like it at all, Parthenon or not. I suggest
those who champion unbridled democracy take a visit to the
Selma shopping center, where market capitalism and unfettered
democracy display their best expression in the entire two-and-
a-half-millennia history of Western culture. Amid the franchises
and discount warehouses is the greatest testament to the American genius and its ability to make life livable even as it destroys
what it is for. Euripides, more so than Sophocles, appeals to us
moderns, that rare talent who knew what a third generation of a
culture was up against, what man turns to next after he is fed and
safe and smug, what evil goes on inside rather than outside a
man's brain.

The sophisticate inside the city walls thus says there is no tragedy
in such magnificent folly, in the grandeur of hitting the wall. Get a
life and find the courage to change, to learn that trees and vines are
ultimately mere trees and vines in an otherwise rather wide, exciting, and complex world for anyone with enough imagination to
plunge in. That loyalty to a dead regimen is only obstinacy—and
at its worst, stupidity, worse when it takes others down with you.
No, the real challenge in this world is nuance and contradiction
within the energy of the town: the struggle between sophism and
ethics, the quandary as to whether book learning is either moral
or necessary, whether culture is ever possible when we are equal,

the role between the rational and irrational mind—the life of the mind and the delights of the sensual that take over once the hands are free to turn the page, not the steering wheel of the tractor. We of the country out here are too few, live too far apart, seek out like kind far too rarely, and are thus often absent from the pages of history and understandably so. We are merely the invisible cement foundation, the unseen piers that go deep into the earth to support the glittering edifice above, where all commerce, art, and culture is to take place.

That ancient divide between city and countryside is still known to everyone born on a farm, who lives on a farm, and who acknowledges that he will die on the farm—and who has lived for a while in a city. The trick again, I suppose, is balance, something between extremes of the rustic Italians Baucis and Philemon eating on wooden bowls, and Petronius' wine-soaked Neapolitan Trimalchio downing stuffed thrush while farting on a silver chamber pot.

The rural man is unhappy in the city, not because he finds it ugly (many urban centers in fact outshine the bleakness of the farm) or because he can either bowl, buy a drink, dance, fornicate, or visit the opera at will. I think it is a more elemental challenge to leave the farm. The denizen of the vineyard or orchard lives in a world of unconditionals: trees must be pruned or there is no fruit; should vines miss an irrigation, they drop berries; an absent worm spray gives rise to defoliation. Man fights nature and there arise clear, easily observable choices. Critical to this fight is the use of muscle. The abstract becomes flesh only through the right arm or left leg. Bolts are tightened, tree limbs sawed, concrete patched, all through the hands and back. Every idea is matched with the flesh. For the farmer, then, there is no conception that is not realized in the physical world, no idea that is not carried to its ultimate and often tragic conclusion. Such an easily identifiable struggle! Such a wondrous thing, when you, alone, with your muscle and your mind in tandem, with or without sufficient courage and endurance, can at least once still, just once yourself, succeed or fail! Such utter and dangerous selfishness!

He who engages in this business for any length of time—or worse, is born into it—can never be happy in town, never be satisfied with a world devoted solely to the mind or human cooperation and construction apart from the solitary effort to harness nature. In meetings, conferences, and lectures, he becomes fidgety. He seeks to halt the speaker rudely if need be, and prematurely to demand someone at least move now! He interrupts the more sober and judicious—as if he wants in this carpeted and plastic conference room to plant an orchard, harrow a row, see, feel, smell something he has created out of nothing. No wonder Aristophanes and Theophrastus found the *agroikos* a raucous loudmouth, a loner who believed that by sheer force of will he could overcome the mob.

The farmer in town becomes disconsolate, even feels subservient, when there is no directive for action, feels even the tissue on his arms to sag through disuse, this boor who sniffs for the direction of the wind, whose eye catches the color of the leaves and the track of the clouds. He interrupts the dutiful consultant and the polite workshop leader and rudely—and I think rather insanely—demands, "Just who is going to pay for all this?" or "Now tell us just what is your role in all of this?" ending with "What do you get out of it?" The utter arrogance of this farmer who feels he can organize society as he has so selfishly his own southwest twenty! I have seen the cruel and arrogant farmer smile when he at last understands why the county, the corporation, the labor organizer, the power company, the broker—all of them!—are at war with him, all of them envying and hating him at each moment of every encounter. In other words, this last generation of family farmers in America is not modest. The agrarian still transfers the solitary world of plant mastery to the human community, and almost for the first time in his life discovers a beauty in his now newly discovered antithetical existence—one unbeknownst to him until he saunters past the city limits. Mr. Yeoman realizes abstractly now that if he is not at odds with the consultant and workshop leader, he knows something is very wrong within himself. He is not as those in town. But instead of being embarrassed by that fact, the farmer, if only for a time, now grows rather to like it! Worse still, he almost invariably loses, knows he loses—and prefers gladly, arrogantly, to lose his way, rather than to win theirs. Or so it has been until now.

But the sober farmer in his doomed exaltation is not foolish enough to ignore that it is precisely the town that has given him the knowledge of being another. Without a community, his hard-won knowledge of the *polis* is lost and without transference. Without such a community nearby, he is nothing, a savage frontiersman, not a builder of families. The farmer needs the town, and the town him: too much of one or none of the other, and both perish. The grower needs the eater, the fabricator, the supplier of his shovel and tractor, the other world to define him by what he is not. The farmer—thanks to the town—now learns, after all, that he is tiller, mechanic, builder, grower: the autarchic man who is everything because he can do a little of everything.

The county planner, the environmental control officer, the agricultural extension agent, the pesticide regulator, the sheriff, the labor inspector, they are all hardworking and often courageous ants who ensure our comfort, but whose existence can unfortunately cease once the central nervous system of the colony falters. They do provide greater abundance and affluence, leisure too, through—as the touched Plato advocated—economy of scale and specialization, but at the price only of bothersome complexity and eventual fragility. In the wild, the farmer sees the wages of their devotion and subservience. The bees left behind when the hives are taken away die without a queen, without the day's orders and directions. Their overpaid, apparatchik administrators alone really know what they are to do. On their own, the queenless colony's remnants rather quickly litter the ground. Even the lowly and rather dullard worm has a better chance than orphaned bees to live one more day. The farmer is there, then, to teach those who are of the hive of their vulnerability, and to provide them with their humility.

In the eyes of the urban bureaucracy, which characterizes most life in the city now—be it school, business, or government—man can also hide. His failure to produce is no longer much a question of his muscle, much less his audacity. I have seen burly Sheetrock installers and stalwart teachers who were fired for reasons other than their ample biceps and excess of spirit and skill. On the other hand, when the truly mediocre urban man files the wrong report, teaches a bad class, measures and saws wrongly, misspeaks before peers, misreads his meter, it is the failure of boredom and rote, a shortcoming that often has few consequences among sympathetic and indifferent auditors or even envious detractors in their midst. Unlike the countryside, the new city can take the muscles off a man's chest and the courage from his heart.

There is beauty for the farmer in being in the countryside, when the country is defined as not being in, but near, a city. But once more, note the importance of the city. Without the city, the farm is nowhere and is usually not beautiful. Without a nearby town, the farmer is in a desert, up on a mountain, a hermit in the wilderness. The farm might as well be on Pluto. Yet put a town nearby, and the agrarian discovers beauty in being what the city is not. He realizes what he is only by recognition of what he is not. Without a town close by, the farm and farmer have only nature as their yardstick. They can turn out to be not much better than hunters, gatherers, or trappers without civilization to which they are integral. As Virgil taught us in his *Georgics*, the frontiersman and nomad have no purpose, no one to feed, no one to share the wisdom of cultivation. Mr. Crèvecoeur agreed:

> Of all animals that live on the surface of this planet, what is man when no longer connected with society, or when he finds himself surrounded by a convulsed and a half-dissolved one? He cannot live in solitude; he must belong to some community bound by some ties, however imperfect. Men mutually support and add to the boldness and confidence of each other; the weakness of each is strengthened by the force of the whole.

The orchard is not the forest or the prairie; the farmer knows that instinctively. His orchard and tractor, it is clear, are not the deer or the hawk. But the farmer learns only from town that he is also not quite the shopkeeper or the banker, and his farm is not the factory or the store. Vast estates without towns are not civilization, nor is culture a metropolis without farms nearby. Neither the empty plains of Thessaly nor bustling Alexandria gave us Homer, Pindar, or Thucydides. The farmer, who is so often solitary, survives because others inside the city walls want his produce; spiritually, he continues because there are those in town he is both needed by and different from.

When I drive now through the West Side of central California, where small towns are rare and nothing more than quarters for the working poor, the land seems barren. Even the air is unfruitful, despite the bumper crops of cotton, rice, and alfalfa. You see, no real town is nearby, the urban manifestation that defines by contrast the farmer's patchwork of trees and vines. Over there on the corporate side without cities, the farms seem ugly for their absence. The overseers—suburbanites from Fresno—in new pickups look not like farmers. Not one of these mid-level employees would ever live where he works, send his kid to school next to his company's cotton field, or have his hired minions over to share a beer on his porch in town. It is ancient Messenia of the helots, not Boeotia or Attica.

Among the latifundia without communities, how do the owners of such barren expanse know that they are farmers? Or rather, without small towns amid their land, how do these farmers at the millennium know that they are not really farmers at all? And how do they know that without towns their cultivated lands are not farms? And where and what are suburbanites, their tract houses distant from downtown and yet islands between doomed vineyards and peaches? Hell will not be the blacktopping of America, but the sort of blacktopping of America, the situation in which we cannot distinguish farmland from suburbs. That is what is at stake with the loss of family farms: not food, not security, but the countryside whose culture created America and from time to time knocks it back to its senses.

THE MYTHOLOGIES OF FARMING

(For Those Who Would Like to Like Farmers)

He sees not that sea of trouble, of labour, and expense which have been lavished on this farm. He forgets the fortitude, and the regrets.

—J. Hector St. John de Crèvecoeur,
Sketches of Eighteenth-Century America

All this happened on a single day one week.

I opened the mailbox and flipped to a random page of an advance copy of a book on farming to find: "The natural serenity of the farm . . ."

The phone rang and a kind voice said: "You farmers are the nicest bunch of people in this country."

An acquaintance from the campus greeted me: "You're so lucky to live out there where everything is so simple."

On the television blared an empathetic head: "Will we still have food once our family farms are gone?"

In a magazine a sensitive writer expounded: "Farming, the oldest and most timeless of man's activities . . ."

Elsewhere, I have combated the mythology of the violent, illiberal, and uncouth agrarian; now let me refute the five commonly held myths listed above by suggesting that the opposite image of a kind, simple, and gentle agriculturist is equally untrue. Contrary to my inclination, but by necessity, I must define farmers as less admirable than the fantasies of nonfarmers.

Myth I. Farming Is "Serene"

In some sense farming is peaceful—out here there are never traffic jams, few people, and not much noise in comparison to the city. Murder and rape are less frequent than in the country. And there is no X-rated theater, crack house, or all-night hotel outside my window. We hear the sirens from town, not vice versa. Many of the greatest philosophers in the West have noted that rustic morality stems from the simple absence of temptation.

Yet Pax Agraria is a myth. The farm as a tranquil abode is the dividend of our romantic and pastoral traditions that date back to third-century B.C. Alexandria, where sophisticated and citified Greeklings dreamed that they were shepherds in Arcadia. Trapped in concrete, asphalt, and stucco, urban man idealizes— the academics would say "constructs"—what he does not know but wishes to be true, as either hope or penance for his own some-times unsatisfying existence. So farms become "serene" and "peaceful" for those dreamers, who under no circumstances would live there. The idea of the calm North Forty is part of the same romance that explains why city folks buy enormous and awkward four-wheel-drive Sports Utility Vehicles for rush hour traffic, or wear heavy, uncomfortable, and treaded high-top work boots just to navigate over carpet and tile. Equipped with such ap-purtenances, they can travel anywhere and so go nowhere. I sup-pose Plato would say that their reason and appetites are not rural, but their suppressed spirit is—the third great portion of our exis-tence that longs for something primordial.

In reality, agriculture is frantic. It has cacophony and a frenzy as breakneck as any I have seen in town. Consider, for example, not the busy harvest or preharvest, but the month of February, Virgil's purported dormant "off-season." Then farmers should be in by the fire, waiting idly for their vines to reawaken, quietly whittling to the hushed rhythm of a somnolent nature.

More likely the following is the winter vineyard scenario.

Pruning is now almost finished. But you can't just tie the selected vine canes back on the wire. Why? Because the wire has been cut, the stake staples torn out, a few stakes crushed, even a few end posts (which anchor the wire) at row's end broken through 365 days of use. Indeed, sometimes the wire of the whole row is on the ground. Pruners are paid by the vine, and so they do cut the wire in their haste to surpass the minimum wage. And they do pull grape stakes down as they yank recklessly on stubborn canes. For the prior eleven months tractors have hit posts, stakes, and vines—you now discover all this flotsam when the vine leaves and brush are gone and the year's detritus of the vineyard becomes clear at last—and for a moment.

This wreckage has to be cleaned up before buds break out in a few weeks.

So immediately after the vine is pruned, you now madly begin to patch wire, replace stakes, end posts, staples—all in order to send men back through to tie vine canes on the wire.

Remember that pruners have thrown their cut brush in every other vine row. But Hank Ortiz and his thirty-year-old brush-shredder are nowhere to be found—his salary is to be made solely in the month of February, so he is custom-shredding 5,000 acres of vineyard too many. He promises only that he and his motor home will be in your yard at 2 A.M. and out by noon. No Hank Ortiz, no shredded brush. No shredded brush, no clear vine row; no clean vine row, no cultivation. No cultivation, no furrows. No furrows, no irrigation. No irrigation and the vines are naked to the frost. Frost and no grapes.

And it's now but twenty days to bud break.

But you forget for a time about Hank Ortiz, because you also need your berms weeded; in a week or two broadleafs are up. The taller those weeds grow, the harder they are to kill. Either plough them out of the row or spray them. But you can do neither without the canes tied; otherwise they flap into the sprayer's mist or the plough blade. So you fix your wire, stakes, posts, so you can tie your canes, so you can spray and cultivate and shred.

But it is now ten days to bud break.

Forget about weeds and worry instead about fertilizing, so that nitrogen is in the ground and ready for the vines right after bud break. Yet to run the fertilizer rig, the vine rows should be weed-free—and the brush shredded and the canes tied. You intend your fertilizer rig to whizz down unobstructed vine rows, the dirt hard and clean for the shanks behind. Yet the fertilizer rig is rented out elsewhere. And the brush is not shredded. And the wire is not yet patched. And the canes are not yet tied.

And bud break is two weeks early this spring.

But stakes, wire, spray, fertilization, shredding of the vineyard are nothing to the orchard across the alleyway, which is now near blossom. Trees have a more complicated dormant sequence of pruning, shredding, fertilizing, spraying, and bees; once their cycle is disrupted by this Mr. Hank Ortiz and his now nonexistent shredding machine, everything for the next year goes wrong.

The problem? Irrigated Mediterranean agriculture, as the historian Mr. Braudel wrote, is always a race. In such temperate climates, the dormant season between leaf fall and bud break is only three months. All the orchard's pruning, the vineyard's tying, fertilization, dormant spraying, weed control, trellis maintenance, grafting, and replanting must be done in that tiny window of ninety days, when there are no leaves. In Mediterranean agriculture there is not much time to clean up the postharvest mess in time for bud break and the long year to begin anew.

Again, the rub? Many of these tasks are sequential and dependent on another, and so cannot be accomplished out of synchronization and without the proper prerequisite job. Canes are tied only on wire that has been fixed; tractors drive only down vine rows that are free of brush; bees are put only in orchards that have not recently been sprayed for weeks; oil spray cannot go on popping buds or leaves.

Usually these frantic steps could be accomplished if the farmer were on his own and in control. He is not—not ever. His autonomous world is not so autonomous, but rather predicated on shredders, beekeepers, pruners, tiers, sprayers, and a host of other freelance workers whose success or failure depends solely on how many farms they can serve before their tardiness and missed appointments result in a confirmed reputation of unreliability and thus termination. The trick for the Hank Ortizes of the farming cosmos is to sign on for as much work as in theory might be done under perfect circumstances.

In winter, circumstances are never perfect. It is foggy here in the Valley in winter. Trucks get lost and are wrecked. It is rainy here. Fields get muddy and cannot be entered, by man or machine. It is sometimes hot in late February here, so vine and tree buds pop and swell weeks before they are supposed to. There are people who do not farm here. Thieves, neighbors, and others get in our way and we in theirs; permits, releases, and paperwork are not always easily obtained from those who are on another, a bureaucratic rather than natural, schedule. No, the farm, even in its quietest month, is not serene.

I will pass on the ninety days of "peaceful" dormancy needed to ready the other twenty-seven tree and vine crops on this farm.

And I will spare you, reader, a whine about the tractor engines to be rebuilt, the shed to be fixed, the pump to be pulled, and all the other tools and appurtenances that, like their natural counterparts, are but resting for ninety days for their needed maintenance and attention.

Farming is hectic, not peaceful at all. I have lived in Santa Cruz, New Haven, Palo Alto, and Athens. While it was loud and brutal in those cities, I sat rather slothful in my relatively quiet and comfortable apartments, always thinking of all the farmers back in Selma in their purportedly serene winter frenzy.

Myth II. Farmers Are "Nice"

This is the old myth of the noble savage, which grew from the Romantic counterattack against the dry and artificial world of the European Enlightenment. To those in suits and ties, in office boxes and on smoggy freeways, farmers are to be aboriginal creatures whose muscles force the earth to give forth its bounty. Like the animals they live and work with, agrarians must be simple, hardworking brutes who, freed of urban stress and gratuitous insults and violence, are as one-dimensionally kind as their environment. The farmer, free of the city, reverts to a natural cycle, which returns him to his prestate self. Thus he is purportedly "nice." No wonder French farmers have it so good: they live in a culture that never really freed itself from the silly romance of Rousseau.

Hardworking? Yes, farmers are—almost all of them. Honest, dutiful, law-abiding, and moral? Yes, as a rule, they tend to be that as well. Eccentric, occasionally stubborn, sometimes adamant? That too is a fair generalization, and they can be near delusional as well, as the price of their isolation and solitary existence and their childish trust in the next year's deliverance.

But nice or pleasant? Hardly at all. Even our first agrarian propagandist, Mr. Crèvecoeur, came close to confessing the truth of the farmer: "If his manners are not refined, at least they are rendered simple and inoffensive by tilling the earth." That is the circular *mea culpa* of everyone in this family: we as farmers are rude, but it is OK because we at least farm, which makes us rude.

In this age of therapeutics and victim-obsession, I could weave a long exegesis that the uncertainty of raising food, the duplicity involved in modern food distribution and sales, the antagonism of government and corporations, and a host of other -isms and -ologies all explain our rudeness and indifference to deportment. After all, in America it is hard now to work for 365 days on borrowed money in the hope that someone you have never met— someone cleaner, better dressed, better educated as well—will pay you money that is owed so you can pay back the bank. The job of farming made us unkind, I whine. But who cares about the cause? Farmers have always been impolite and not sweet.

A family member calls, then speaks in monosyllables, and abruptly hangs up. No hello, no good-bye. You object that this curtness is either a result of family intimacy or accepted casual telephone protocol. No. He and all the rest out here are that way to others and in person. A neighbor lady calls up later that evening, "Say, [*no salutation*] your pipeline is leaking onto my vineyard [*in fact, her pipeline was leaking into her vineyard*]. Just thought I'd let you know so you'll fix it tomorrow [*still no identification*]. Good-bye now, this is Hilda Brightwell east of you."

I once planted an orchard that grew no plums and a vineyard whose grapes always rotted. Add to that pears that burned up, quince that died, and nectarines that dried up from nematodes. The neighbors knew that circus better than I. And their response when they saw me in defeat at the property line or on the curbside?

Polite indifference? Feigned ignorance? Sympathy? Never. "Well, I guess you boys planted about a hundred thousand dollars' worth of trouble, now didn't you." Even the kindest commentary went like this: "There's a lot like you who planted those no-good plums, and they got the same nothing as you did."

Even agrarian death is not mourned for long, but is accepted and expected as the natural wage of farming. It is seen as a slight road bump in the way of next year's harvest. Of a recently deceased resident of some thirty years, I have heard the following. "Well, that old bastard just plopped over out there in the field, he did. Kinda liked him too, even when he bragged more than he should. His boys will have a field day dividing up that mess, and his widow better watch out for the no-goods who might like to move into that big house with her. Bet the whole thing's sold and they're all in town and away in a year." *Requiescat in pace.*

You object that these cuts and rebuffs simply illustrate the petty absence of manners. Policemen and stockbrokers do it too. No, again. The absence of affability is more widespread out here. We don't have to be polite to some, rude to others, as the financial situation demands. Instead, we are generally quite fair and honest in being rude to all, we who are kings of our own eroding fiefdoms, blustering Frankish counts safe but surrounded in our castles of the Morea by hordes of Turks. We are not, then, by necessity uncivil to our inferiors and obsequious to our betters, as is the general creed in the corporation or government, as Isocrates said of the Persians. We can afford to be curt with all without prejudice or social bias, without worry about our own futures—with confidence (often misguided) that telling the truth does not harm us economically and brings with it moral reinforcement as well. A man who sees enemies of his plums and peaches on every horizon after a few years rarely smiles.

Yet unniceness at its core reveals a greater paradox in the life of the agrarian—the bitter wages of our bluntness. More than ever, the farmer at the millennium needs the cooperation of like kind to survive and battle the government and, increasingly now, corporate agribusiness. But the very regimen and semiautonomy of the family farmer's daily existence make the rustic uniquely un-suited—though not in theory unwilling—to cooperate, share, and forge any alliance that might save him. Whatever his good intentions, he has not even the veneer of the conciliator. He cannot use diction, dress, or social protocol to mask intent, much less disguise disgust or mitigate the expression of anger. The failed history of American populism and agrarian activism bears that out.

You see, all communal activity is predicated on the currency of simple kindness and good manners. But those are the exact traits that are either unneeded in the farmer's daily solitary existence or seen as liabilities ripe for exploitation by others—or sensed as a bad first step on a long road of monotony and sameness. We farmers apparently do not know how to be nice, even when we wish to be. When you work with dumb plants and animals, there is no reason to be either loquacious or affable. Most of the things we work with—vines, tractors, water, animals—do not have a rational mind. When it is a choice between brutal honesty and euphemism, we choose the former and are rewarded for our truth with oblivion.

I have tried in a small way to organize farmers, gone to cooperative meetings, been engaged in ad hoc and informal attempts to redress agrarian grievances. All have been relative failures—largely because I smiled, shook hands, and tried to appear both mannered and sincere, only to confirm either naiveté or weaknesses or that I had become utterly tame. After the ordeal I earned the loss of agrarian self-respect, which follows from the resort to nuance and subtlety, and the public rebuke of being both the offended and offensive.

A group of us once sued Sun-Maid Growers of California for its failure to return the retained moneys to us, its membership. We held meetings. We sent fliers. We made calls. We spent hundreds of hours reading briefs, plotting strategy, and appearing in court. The hardest part? Dealing with the corrupt legal class? Navigating through the labyrinth of the American court system? Facing the capital and administrative hordes of a huge, inefficient, and mostly godless cooperative? No. It was talking to fellow farmers whom we wished to help and to organize so that we could petition, sue if need be, for expropriated property. Their responses to our communitarian efforts to reclaim their lost capital?

"Now, what you are doing is just fine with me, but just exactly what do you get out of it?"

Or "Why should I put up any money to sue anybody, when there are others with more than me?"

Or "Sure, the whole bunch is crooked, but let me tell you first . . . " (twenty minutes of narration about feuds with his neighbor follow).

Or "Mr. Hanson, just tell me right now: in two years will I or will I not get my money back? Right now, answer yes or no and then shut up and sit down."

Or "The problem is that our damn lawyers are worse than the crooks who took our money in the first place."

And they are probably right. Still, taciturn farmers might have functioned well in an old republic of like kind, but in the democracy of the modern age they really do appear unmannerly. If I wish to be flattered, entertained, treated with perfunctory respect, and met with pro forma chitchat, I will from now on seek out peach brokers or Sun-Maid Growers of California's roving sales agents, not yeomen. If I wish to hear pleasantries, I will not go to those farmers milling around the barn or communal irrigation gate, but to a bank which is about to call in a loan or foreclose on a farm. As a rule, farmers are sincere and they are honest. But they really are not polite people.

Myth III. Farming Is "Simple"

Our third myth is a product of polite condescension and also part of the age-old antitheses between city and country discussed in Letter Five. Anyone who lives in the country, works with his hands, gets dirty, produces something that can be handled and felt, must be rather simple and his world equally so. This fallacy that farming is easy is held not by the mean or duplicitous but by the well-meaning and ignorant who confuse the outdoors with purity and plainness.

But family farming—not corporate agribusiness—is complex, maddening, and inexplicable. The farmer, unlike the agribusiness specialist (who really is simple), fights a war on all fronts. His money? Now an accountant, he figures interest, dividend, depreciation, profit, investment, and costs to beg for someone else's capital to pay for an entire year of the unknown before he harvests. Whatever his physical prowess in the rather mean world of his vineyard, the power of his arms and back cut no ice with a junior loan officer with red suspenders, pin-striped shirt, and round, tortoiseshell glasses. Despite his cut hands and creased flesh, the farmer must battle rather impressive gladiators in someone else's arena, constantly pitted against B.S.'s, M.B.A.'s, and C.P.A.'s whose nets and tridents are spread sheets and software. And they really do and must say, "Mr. Hanson, are you sure of a six percent return on those plums' initial capital investment within three years? Our figures suggest otherwise." *Ave, Caesar, morituri te salutant.*

Surrounded by engines, gears, and bearings, the farmer must be a grimy shade-tree mechanic who welds, fabricates, and changes enormous water- and air-filled treaded tires that weigh far more than he. Some days he disks sixty acres of vines; but on others only two or three as in defeat he welds his broken disk, changes his worn shaft bearings, and rebuilds his ruined alternator. Of the farmer's makeshift tractor rewiring, the local dealer's authorized mechanic—certified at the Ford training center itself—scoffs, "This is a hell of a mess, even if it has worked this long. We would never do anything like this."

Our agrarian at times is no more than a thug himself, a centurion of Caesar's crack Tenth Legion whose task is to cross over the Rhine and battle the woolly Germans. In a world of macho pruners, grimy pickers, and rather angry hired shovelers, polite rebuke—even pleasant compliment—hold little sway. More likely, kindness leads to complacency and soon on to contempt, ending in outright defiance. Our farmer by bluff, brag, or muscle must at times stare down, push, threaten, hit, and run off those of America's forgotten classes who would like to see him try. He must lay down his checkbook, wrench, sugar-tester, and polite technology to wade into a crew to announce to men who are usually not announced to, "The next sonofabitch that leaves only three canes on my vines is fired, pronto." *De l'audace, encore l'audace, toujours de l'audace!*

Puffed up with the notion that he is autonomous and a small businessman of some acumen, more often the farmer is an unthinking brute. In mid-September, the temperature over 106 degrees, most of his raisin-rolling crew suddenly vanished, he must gather up ten stragglers and himself lead these sunstroked men down the

vine row in the scorching heat, shoving the caramelizing raisins—
his entire pay for the year—under the vine, yelling like Frederick
to his exhausted but now hesitating comrades, "Rascals, would
you live forever?"

He himself must drive the tractor down his rows twelve hours
and more in July, fever or not. Some April nights he is up running
water for a week straight. In January he prunes a hundred vines
up, a hundred vines back, his mind put on ice as legs and arms
work in tandem until dusk. Too often victory or defeat is found
only within his resistance against the elements and monotony, not
just within his IQ or his biceps—or the ideal mean between the
two. Bored silly, without cash, tired, the farmer looks at his un-
ploughed field, his gassed-up tractor in his yard, and his relatively
healthy body, and thinks to himself, "Fifteen hours from now it
will all be done." *Del dicho al hecho hay gran trecho.*

But the agrarian cannot live by endurance, reason, or muscle
alone. The farmer must gamble more than any Las Vegas junkie—
to pick early or late, fast or slow, once or twice through the orchard. He can plant plums, peaches, nectarines, or grapes—on a
hunch that one, none, or all will have a market in five years. With
no salary, health insurance, retirement, disability, or unemploy-
ment insurance, the farmer's entire life turns out to be a wager—
that he will make enough to survive when his body is shot, a small
pile to tide him over should his courage and nerve leave. "If we
had not planted those plums, but peaches instead, same with the
no-good grapes, we'd have fifty thousand in the bank, not a hundred and fifty out at the bank," I once told my wearied brother. He
at once answered me back: "And when peaches were no good,
you'd say the opposite or something else, since who the hell
knows what will happen anyway."

And who does?

And finally the farmer is a dilettante plant pathologist, geneticist, biologist, and chemist. No Nobel Prize winner knows all the intricacies of plant production, how exactly and under what precise conditions plants produce food. No agricultural scientist knows the exact—and relative—contributions of weather, soil, air, cultivation, and water that create fruit. No one knows, but the farmer alone pays. So he wracks his brain, reads his books, does his own ad hoc interviewing and research to discover just how many pounds of nitrogen, how many acre-feet of water, how much brush to leave, spray to put on, spray to leave off, how many plums to thin, grapes to girdle, in order to produce food on a particular soil, in a particular climate, in a particular place, in a particular year—all on the premise that success then demands exact duplication the next season when none of the variables will be the same, that failure under all circumstances must never be repeated though exactly what caused failure in the first place is never known. A whole lifetime can be consumed in that.

I've taken meaningless degrees and taught thirty-nine different semester classes from Attic Greek Composition, Roman History, The Origins of War, and Advanced Latin Grammar, to Sophocles, Sallust, and the *Satyricon*. And I have dutifully chaired worthless tenure committees that have never not tenured anyone, created a classics program *ex nihilo* and watched it sputter, helped to raise three kids, written books, remodeled, replumbed, and rewired the house, been married for over twenty years, engineered a ridiculous five-hundred-foot block wall, come down with dysentery in the Valley of the Kings and kidney disease in Athens, researched and pontificated about Thucydides, hoplites, and Diodorus' use of Ephorus, and lived under fascism and then again amid socialism for two years in Greece. I even for a while attempted to lecture fifty times out of state about farms, wars, and the poverty of postmodernism. All that so far has turned out to be mostly free of deadly catastrophe.

But for many years just farming trees and vines in Selma, California? Now that was mostly one big ungodly and embarrassing failure.

You see, it was not simple.

Myth IV. Agribusiness "Threatens Our Food Supply"

Myth IV is the product of the ecological movement. Slightly Marxist, slightly academic, slightly paranoid, slightly scientific, more often well-meaning, idealistic, and utopian, the theory runs like this: Corporate America has now taken over agriculture (mostly true). Those captains of industry see farming only as a business, where profit is the sole arbiter of all farm activity (mostly true). Consequently, they have developed technologies, chemicals, and practices whose single purpose is not cultural, not community-spirited, not ecological, but entirely commercial (mostly true). The result is that we, the people, are being bombarded with food that is dangerous and soon—due either to corporate conspiracy or to the general collapse of their overly sophisticated system—to be in short supply (mostly untrue).

Yet the Truth is, of course, more bothersome than the Lie. Americans pay less for food than any citizenry on earth. Americans receive the safest, least infectious food on the planet, which will not kill the great majority of us. Americans have a greater supply and selection of tasteless fresh produce than any people alive. Those facts are indisputable and are true because, not in spite of, enormous complexes of vertically integrated agribusiness consortiums, whose refrigeration and packing plants, trucking, brokerage, and distribution services manage to navigate harvests across a continent of vast expanse in a matter of hours.

Make no mistake about it: Agribusiness is a godless enterprise. It has created an entire industry to create artificial species of fruits and vegetables—produce that is hard, shiny, colorful, will travel, and tastes awful. It enriches a few at the top, disparages wage labor, contributes nothing to the communities in which it thrives, uses cheap food as a mechanism to consolidate market share, seeks to monopolize farmland, receives free government research and subsidy, is inefficient and propped up by mostly hidden government support, pollutes through bribery and civil and political discourse, and provides the cheap fuel for the entire American materialist rampage.

Without the low-cost, nutritious, and generally safe food that agribusiness ensures, the welfare system would collapse, there would be no Food-4-Less, and McDonald's would go broke. The world of agrarianism come back from the dead would be quite different and—I am afraid, as it was in the past—hated. Food would be local and in season and far more expensive. No growth hormones and regulators. Farmers' markets would be the norm, not the exception. Not just half a parking lot each week, but acres of them every day. Fruits and vegetables would be riper, with fewer chemicals, and therefore uglier and tastier. Suburbanites would find roadside stands every mile; city street corners would have fruit peddlers. In a Vermont market in April, there would not be watermelons, cherries, and apricots from the Imperial Valley. If it froze in Minnesota in late May, there might not be any local apples for the summer to come.

In turn, smelly and unkempt farmers would be ubiquitous, at the restaurant, in the post office, at the bank. And not just in Iowa or Kansas, but farmers in every community and metropolitan district—they might comprise 10 percent or 15 percent of the population, not less than 1 percent. Without the stranglehold of corporate shipping and distribution, local orchardists, vineyardists, nurserymen, and truck farmers would be strapped to supply their communities in season as supply—subject to local weather and harvest conditions—dictated. Local co-ops would process grains, and they would cost more since farmers would control the land, the mill, and the product until it reached you. No more monopolies that put farmers in the street or in the grave. By the 1990s four cereal companies controlled 80 percent of the market; five grain conglomerates ran 96 percent of all of America's wheat exports. All that would go.

All citizens would listen to local weather reports, worried about unseasonable frosts and hails, very concerned that their produce might be interrupted or short—with no Florida or California to bail them out, without tons of surplus kiwis, melons, and nectarines available anytime they wished. Young people would want in on this profession, so central to the community, so profitable to the hardworking, so esteemed by the nation. All sorts of local and antiquated varieties of fruits would reappear: delicious apples and plums with unimpressive hues and crusty veneers, fruits of the nineteenth century that become overripe in a matter of hours, that bruise and discolor when picked, that eat well and ship badly.

Under agrarianism, schools (as in my childhood) might start a week later to allow for harvesttime labor. No more piano, soccer, and ballet after school; no more Jason and Nicole off to gymnastics instead of picking and shoveling next to Grandpappy until dusk. With farms of eighty to a hundred acres everywhere, sons and daughters would tie vines after school; locals without the dole would harvest—and they would pick and prune or not eat. And they would then choose to eat. And so there would be occasional shortages of hands, with no guarantee that five hundred men would arrive on specification from Mexico to pick and then be gone the next day, on to the next corporate enterprise. Chemical use on the farm, of course, would be less frequent, as is true now under the few remaining family enterprises. Poison is used less by those who put it on themselves, who have little money, and who are not part of a chain that ensures people in Philadelphia that their nectarines picked Monday in Selma will look absolutely the same back east on Friday.

In a world without integrated corporate agriculture, without chemical poisons, and without enormous vertically integrated chains of supply and distribution, the produce section of the supermarket would not be open at midnight, and it would not have papayas, guava, bananas, and red grapes in February. There might be three television channels, not five hundred, given a viewing audience in large part exhausted by shoveling until dusk on their small tidy farms. Given the parochialism of local tightfisted agrarians, interstate freeways would not slice through five counties at a crack in a perfectly straight line, and so it might take twenty not six hours to drive from Sacramento to Los Angeles—small plodding produce trucks and smoking pickups clogging the lanes as dirt-poor farmers drove down into the L.A. basin each morning to peddle their wares to a waking and hungry Southern California. We would be a more moral, more law-abiding, and more humane society, but that would be so perhaps because we would have a more exhausted, poorer, and immobile citizenry.

Agribusiness, not family farmers, have given us beautiful and plentiful and bland and tasteless and mostly safe fruit. You find it in any supermarket in America at any time. The world knows this and so copies not our south twenty, with barn, and Gramps on the tractor, but the idea of mile-long rows, and enormous machines, and vertically integrated conglomerates. Agribusiness, not yeomen, makes it possible to have Wheaties in Greece, and Levi's in the Sudan. The Archer Daniels Midland Company and a few like it are probably right in their boasts that they make it wholly probable that 6 billion on the globe can eat cheaply—and so have time and money to experience the banalities of modern culture before they die. This entire life-sustaining multinational enterprise is but one part of a larger seductive appeal to the senses that is surely a world away from the blinkered farmer.

Remember, the family farmer is not even fair or democratic, at least not entirely. While agrarianism functions only within the realm of capitalism, it has always been burdened with an ethical repulsion for the world of commerce; its rote, and tradition, and moral investment do not produce goods and services to the same degree as the corporation. The latter is godless and without memory and not shackled by voices of grandparents in its head—and thus free to lay off, rip out, move on, tear down, or take over as the laws of supply and demand alone dictate. And dictate they must if all of us are to eat and enjoy as we demand.

Confess it, reader: agrarianism come back alive would not welcome such huge corporate chains of food, dry goods, and mass entertainment outlets that give us more than we need at cheap prices we can somehow manage to go into hock for, a small tab really for destroying the smug and oh-so-tiny cosmos of local merchants and century-old craftsmen. A culture of agrarians would be uneasy with the demagogic idea of sweeping entitlements and large government intrusion. In short, let us be still more honest: the family farmer has little that those in America need or want. The world of IBM-cloned computers, CDs, and Disney's *Lion King* is what the planet prefers—and that partiality is ultimately quite democratic and gives much to many on demand.

It is in the interest of corporate America to sell goods to everyone who can obtain credit, to everyone of every color and creed who are united by an entirely color-blind CD player or colorful Gummi Bears, who all alike—and quite democratically—pay the same 16 to 19 percent interest on their overdue VISA accounts. Class, like race, like ethnicity and religion, has at last met the democratic juggernaut of global capitalism.

In this brave new vulgar world—which today's pampered critics never understand is ultimately antihierarchical, anticulture itself, and so purely democratic—agribusiness operates. It, not agrarians, ensures food to millions, thereby saving for the Wilsons, Martinezes, Yangs, and Husseins the worry of finding rice, porridge, or bread at an affordable price. Only that way can they rent movies, buy a plastic Christmas tree, or get braces and eye tucks at affordable prices. Corporations, at least for a while longer, will continue to produce literally tons of food for us from mile-long fields. Those food factories alone allow Americans to buy peaches for 49 cents a pound, fifty-pound bags of rice for a pittance, or cantaloupes three for a dollar at almost anytime and anywhere. Megafarms ensure that tomato paste and soy by-products, cottonseed, cattle offal, pig eyelids, corn syrup, and grape sweeteners are not thrown away, but can be mixed and matched and chemically laden to give us concoctions like Pop-Tarts, Ball Park franks, veggie burgers, and Lucky Charms. Corporations give us vegetables in bright plastic bags and plastic-covered hormone-laden meats that are clean, mostly safe, and bloodless. If it has to feed us, corporate America someday will be able—and quite willing—to recycle our very flesh and bones: corpses freeze-dried, smoked, processed, concentrated, ready to eat or microwaved, hyped on Rush Limbaugh and blared out on MTV, Granny's ears and little Josh's nose ground, puréed, and artificially sweetened into Baby-Bites and Fruits-Are-Us, mad man disease or not.

Family farmers, in contrast, slowly and with conversation, put apples in paper bags, hand corn to shoppers with dirty hands, and have insects crawling around their reusable boxes. Again their fruit looks awful and tastes wonderful. In this world the sellers talk not of price but of how they grow food—the entire boring tale of watering, fertilizing, cultivating, and picking it that you have no time to hear while the kids are fighting in the back of the Explorer and the cell phone is two calls backed up. And family farmers do not worry us with toxic soups or chemical residues, but man's age-old nemesis, the bacterium, is not entirely eradicated from their produce. Farmer's natural milk, unprocessed juice, and chops on a hook are more likely to have *E. coli* and a fly or two.

So give agribusiness its due. Yes, it helped to destroy the agrarian profile. Its onset wiped out thousands of small towns and communities. Corporate farming obliterated the entire rural culture that was once America, and for better or worse, was integral to the appearance of the uniquely American twentieth-century material appetite. It took hardworking, dirtied, and dutiful underpaid sons and daughters off the land and into suits, air-conditioned offices, and real money. Latifundia brought to farming huge, horrific machines, an army of accountants, brokers, and bankers that hated the idea of a bumpkin ploughing on his granddad's twenty. It equated the use of the land with the worst corruption of the human spirit. Corporate agriculture did all that and more still that we will only come to learn of later.

But agribusiness has not yet given America food that is immediately dangerous, scarce, or expensive. It has not yet forced down the gullet of America anything it did not want. Farm preservationists and agrarian activists are right to want the Lie and to hate the Truth. But finally they must be honest and so must acknowledge the Truth: agribusiness is dangerous and frightening, not because it has failed, but because it has succeeded beyond our wildest expectations.

Myth V. Farming Is "Timeless" and "Forever"

We read, see, and hear that agriculture is of great antiquity, a timeless part of the human experience itself. It is not.

Our final myth of agricultural perpetuity also derives from the
back-to-nature movement, or perhaps even from the anthro-
pocentric idea that all of nature exists for and is defined by us. But
man is very old, and nature is older still, and agriculture is very
young—and so there is no reason to believe that farming will always continue as it has or at all. Man has inhabited this planet for
over a million years, *Homo sapiens* perhaps for the last 200,000
seasons. Cultivated crops on any scale and true agriculture have
but a 7,000-to-10,000-year pedigree, one coterminous with civilization itself.

My point? Simply that for most of the life of the human
species, there was no such thing as agriculture, as is true even still
with a few indigenous tribes today. When such a premise is accepted, then its logical corollary is apparent as well: just as agriculture is a relatively recent development, and not essential to
human survival, so too it can disappear and not end man's continuance on the planet.

The link between farming and mankind is not survival, but
rather civilization. Man can live on without agriculture. But civilization, likewise a late and frail phenomenon, cannot. For man to
be stable and fixed, to form populous communities, to have surpluses (Aristotle's material safeguards so necessary for the life of
contemplation and intellect), to be literate and to be lawful, he
must grow food, which in turn ties him to the soil, to one place as
it teaches him what property and culture really are.

To the Greeks, the *polis* was simply a reflection of a stationary
and landed populace who grew food and planted permanent
crops—in antithesis to Scythians, Thracians, and other nomads
who hunted, fished, raided, plundered, trapped, and traded, but
did not have a permanent agriculture and thus no civilization as
the Greeks knew it. No wonder that the Cyclopses, Satyrs, Centaurs, Amazons, and all the other monsters of Greek mythology

are creatures of lawless disorder who have one common and feral bond: they do not farm. No wonder that when Odysseus meets Polyphemus, Calypso, Circe, the Lotus Eaters, Scylla and Charybdis, he meets humanlike creatures of assorted shapes and sizes who as nonfarmers are kindred in their uncivilized states. No wonder that when Thucydides at the beginning of his monumental history wishes to emphasize the barbarity of early pre-*polis* Greek civilization, he says merely that "they planted no trees or vines." His readers, of course, would have nodded in their agreement. Farming, then, arose late in human history, and with it civilization. But will it always continue?

Of course, there are apocalyptic scenarios for agricultural demise—nuclear exchange on a global level, chemical pollution of the atmosphere, or epidemics of strange new plant viruses. Alarmists, in theory, could be right that productive but genetically vulnerable hybrid species, together with a growing dependence on a few technicians and petrochemicals, could make food factories vulnerable in the next century to such unforeseen challenges. We could also soon die prematurely at middle age due to chemically laden food. But it is unlikely that food production itself will cease and that mankind, in vastly reduced numbers, will be forced to return to our pre-state origins as a roving society of a few thousand hunters and gatherers.

Far more likely, agribusiness will thrive and thus ensure that food-making in some form will continue even as actual knowledge of farming exists among fewer and fewer people. Even now in this country no more than a million or so Americans if turned loose know how to produce enough food to feed their peers at present levels of population and material comfort. If our past is any guide to the American character, it is likely that corporate enterprises of the next century will be devoted to creating even more food—will that be the proper word for it?—under more efficient and sure circumstances: meaning less human or animal involvement, and more predictability in a realm beyond soil, weather, or muscular labor. The industrial science of mass fabrication of food, which has a pedigree of only a few decades, will accelerate, but the age-old craft of what the Romans called *agricultura* will erode. Agriculture, after all, means not food production, but "the culture of the soil." And I see no assurance that in the millennium to come food-producing plants will be grown in the "soil" or that there will even be such plants, much less their "culture"—much less farmhouses, rural communities, and families tied to particular parcels of land. With the disappearance of this culture, the question then arises: will there be civilization as well?

Yes, there probably will be a complexity in the sense of a sophisticated urban landscape and a specialized workforce. But will it be a civilization we are proud of?

The hour is late for the American farmer, and what is needed is not more romance and mythology but a truth that is often brutal and offers little comfort to anyone. Farming, always difficult, dirty, and sometimes deadly, is now even more so, given years of static commodity prices: farmers must work harder for less money, and they feel and show that struggle. Farmers, always curt and blunt, due to their solitary existence, are even more so now, given their vanishing numbers and the truth that they are failing and going broke as millions of Americans thrive as never before. Farmers, always harried and versatile, are even more so now, as they are forced to be accountants, tax specialists, mechanics, computer-literate, and conversant with zoning and environmental laws, as well as age-old growers of food.

In contrast, agribusiness is not more complicated, not more vulnerable, and not more at odds with contemporary America than are family farmers. The corporatization of food is simple and operates on a single truth: there is no money in growing harvests, but a great deal in packaging them, shipping them, and selling them. Invulnerable is the conglomerate that can do all three, which can lose money from growing food and profit enormously on getting it to you. Its upper- and mid-level employees, with health benefits, retirement plans, and usually clean and comfortable workplaces, are polite, kind, and relaxed folk more to America's tastes than farmers, who have no money or time for such things. They have been busy, we must remember, going broke growing only food.

It is easy—and becoming an unconscious and natural part of the American character—to develop housing tracts from farmland, to shop at the supermarket produce section, and to eat anything at any time while romanticizing from afar the man with the hoe; but it is hard to curb our appetites, buy direct and in season, and keep the countryside pristine to benefit someone illiberal and bothersome in our midst. We wish to make the farmer like the suburbanite in appearance, behavior, and ideology when he is assuredly not; and then, and only then, save him on the cheap and in the abstract. But far more honest and difficult it would be to confess the truth about his nature and then rescue him in the concrete.

Family farmers are not more moral than people in town; and what they do is no longer essential to the life of the nation. America will continue to be free, rich, and democratic long after they are gone. But they are different and they are a link with America's past that brought us the very bounty we take for granted, or worse, sometimes despise. You off the farm are not truthful in claiming farmers to be saintly and invaluable; and we are more dishonest than you for basking in that romance. You in town should like us for offending, not pleasing, you; for not wishing to be like you; for that is ultimately for your own good. And you farmers, as you vanish, should not claim that you are not disappearing, or that your disappearance will destroy what your country has become. If anything, you should cease your mythmaking and must feel proud, not ashamed, that you are bothersome, direct, unchanging, and so, in your eleventh hour, entirely and forever at odds with all that which you are not.

Part Three

MAN VERSUS SELF

LETTER SEVEN

AUTOLYSIS

(For Those Who Think the Farmer Is Sometimes
His Own Worst Enemy)

*No imagination can conceive, no tongue can describe . . . the man
of sorrows.*

—J. Hector St. John de Crèvecoeur,
Sketches of Eighteenth-Century America

A great foe of the farmer, perhaps the most prominent of all, is not
others—man, plant, or animal—but himself, the agrarian cannibal.
The farmer does more damage to his trees and vines than any rabbit, bug, or weed. I do not mean here the farmer's propensity to
maim and wound himself. Agriculture is America's most dangerous profession, and a few who regularly venture on this farm are
without right arms and are absent an eye or ear, their flesh burned
or replete with scar and scab. A few of us ourselves—brother,
cousin, father, and self—have been run over, poisoned, run
through, and lost fingers be it by tractor tire, spray rig, or hay rake.
When you are alone, in a hurry, without consultation, and among
blades and gears, you pay with your flesh for carelessness.

I also do not mean in the abstract our own propensity to produce to surfeit. Our habit is to go it alone, to grow what is not
needed, to call communist any who say unite and organize. Whole
books have been written of the blinkered agrarian whose own

stubbornness, so necessary for survival, is what dooms him. If only he could, as GM or Ford, shut down his production line for a year or two, people, now hungry, would give him his due and learn to value what they now do not have. Farmers' own creed and behavior have sentenced them to produce to excess even as surplus is destroying them. Visit a meeting of farmers outraged over bad commodity prices and broker theft (always impromptu and ad hoc): the level of analysis is brilliant, the exegesis of cause and effect prescient, and the final call for sustained organization and action inevitably disappointing. Apparently such meetings are opportunities for the brooding and sullen farmers to leave their tractors, scream to one another that the grower, in fact, is not dumb but knows precisely what chicanery is being done to him and why. And after that public and group catharsis, the farmer, now content, mopes home to suffer and die in private, convinced at least that his gravestone will note that he did not perish in ignorance. He is content in his doom that at least he took no handout, wanted no complicated and rather boring master plan of salvation, replete with lawsuits, consultants, and do-gooders, who this time—but not next?—were on his side. Yes, farmers go to meetings to scream about the injustice of commodity pricing, then at the lunch break trade insights and lore about how to squeeze more plums, grapes, and peaches out of their own particular near-bankrupt farms.

Rather, I mean here that the yeoman's real lethality derives not from his temperament or national character, but ultimately from the sheer concept of scale. Remember, he is no backyard gardener who errs on a fertilization rate and kills his two prize rosebushes with too much Miracle-Gro. Make a mistake on the farm and it is repeated a hundred-, even a thousand-fold through the grid of the orchard and vineyard.

Arnold Peeks, resident farming genius of these environs, once
went to Italy not to see *David* or *The Last Supper,* but to view the
vineyards of Tuscany. He came back starry-eyed at southern Europe's beautiful arbors, terraced hillsides, the dividend of exotic
agrarian protocol that promised a novel, a better, way of farming
than our boring practice near the now pedestrian Fresno. What
else but misunderstood and misapplied foreign vineyard technique would explain why Arnold Peeks subsequently stripped off
the bark of every one of his 20,000 vines, coated the cambium
layer with oil, and felt he at last, like the Italians, had eliminated
the only habitat of the worm and bug that crawled from soil, to
stump, to his grapes. A chemical barricade now blocked the
crawlers' highway to his fruit, he reasoned, once the rich habitat of
woody stump had been replaced by noxious chemical slime.

Had he invested one year and one vine to such an experiment,
he would have learned that the oil—his own lethal concoction, as
it turned out—would penetrate the circulatory system of all his
vines. It would make its way through the canes and leaves, and so
sicken and discolor thousands of bunches, before ultimately sickening the vineyard itself. The oil, his oil, his oily idea, accomplished that annihilation far more efficiently than any rabbit or
mite, better even than the thug with chain saw. "Why, Arnold
Peeks, they say, ruined his entire goddamn vineyard, ruined it
himself even, just look at it, it about makes you sick." Rubber-
neckers in their pickups gawked as they cruised by the carnage.
"Why, he cut off his own bark and oiled it up good. Why'd he do a
fool thing like that?"

But let us be fair to the Arnold Peekses of this Valley. Because
agriculture is largely a solitary existence, demons race through the
farmer's brain—weird ideas about how to do something better,
cheaper still. It is a strange hubris of ours to think that the unfathomable processes of the plant kingdom are knowable after all to the

viticulturist—if he will only be unafraid to dare. Rarely is the farmer talked out of his grand designs. Far too few exist to debate and check his folly. He is the remorseful and the re-actor who learns after, not before, the consequences of his impulse. He is something like the writer, who crafts entire worlds inside his brain, convinced they are logical and of interest to his fellowman when they are neither and in fact often either incomprehensible or absurd. Let us then empathize with Arnold Peeks and his ambition.

On one July day, I debated over a final sulfuring of the vineyard: On the pro side, it was insurance against a last gasp of mildew before the grapes sugared, a sure way to end the season with a clean vineyard. On the contrary, it was well over 100 degrees and rising, nearing the temperature when sulfur burned rather than fumed, scorched rather than protected leaves. Still, the thought of ten hours of burdensome drudgery tipped the scale toward petty catastrophe. After all, the workaholic farmer who simplistically judges all by their willingness to be mobile reasons always that the greater danger is always inaction rather than action. "Nothing bad can happen if only a man works hard and honestly" is what I heard my entire life—so for ten hours I gladly opted for the sulfur. I was convinced that the toil would result in not a single bunch lost to mildew.

I was right. Two days later those vines burned far better than from any spider attack. Far uglier they were than the parched vineyard of the slothful farmer, who will not irrigate his crop so that he might plunge into the lake high in the Sierra. It would have been better all around on this farm that July day had I floated on a tire with a beer on my belly among the lazy in the mountain lake above.

Because man possesses intellect and not merely has the power of his own muscle and brain but access to machines and capital, his damage can be really quite lethal when self-inflicted. Who is to stop the do-gooder farmer at work from destroying his crop? What gun is to be put under his nose, what poison to short-circuit his nerves, what fence is there to keep him out when and where he should not be, this poor, well-meaning, but now dumb farmer, so intent, so smug, so hardworking in his efforts to destroy himself with fertilizers, chemicals, and strange concoctions? Farmers, you really cannot find the divine secret of how to grow perfect grapes; give up that doomed quest before your search destroys your entire crop.

Farmers, I have learned, do not believe psychology exists. It is but a nineteenth-century creation, like phrenology, at worst a pseudoscience that reasons evil is but sickness. At best, the formal study of the mind is a trumped-up branch of talking, just maybe counseling, or perhaps what the Greeks might call a tiny subspecies of literature and philosophy. Yet, if not part of psychology, what is this inner fight against oneself, this too common practice of the farmer working to destroy himself? "Yes, I left my trays open to the rain when I should have rolled them," the neighbor Vaughn Kalderian told me, "but you boys are four together, and have somebody to talk it out with. But me? I had too many voices in my head to know what the hell to do this morning. So I left 'em open, let 'em rot right before my eyes, I did."

"Yeah, Vaughn," I answered. "We can see them rotting right now. You just let them rot, it looks like."

When you burn your vines with sulfur, when you thin your plum trees before a natural fruit drop, and when you leave your raisins naked to the rain, the true enemy of man appears in the mirror. But the autocide becomes real and lethal not merely when he seeks to do himself and his kin harm, through either drug, drink, or the fist. No, just as often—more so, the farmer swears— we hurt ourselves and our own when we wish to do too much good, when it would have been better for all had we long ago ceased our efforts. I finally learned at thirty what my grandfather had told me at ten: "Let it alone," he said, "don't work it, you'll only make it worse than it already is." Again, Plato was right about man's tripartite existence: he is governed by his appetites, and by his reason—and by his own indomitable and sometimes unstoppable Will. A free man, with brute experience and acumen, without overseers, and with a creed that he must act, not merely talk, surrounded by lethal machines and chemical concoctions, at times can be a very dangerous enemy of himself.

True, this enemy of agriculture, ourselves, is not the most lethal adversary. But it is the most difficult to keep at bay and it does the most spiritual damage because it eats at the heart of the farmer, when he does wrong in his attempt to do right, and when he, not others, has done himself wrong. Parents and benefactors and all you civic-minded, remember this from the vanishing farmer, remember how hard it is to keep the tractor in the barn sometimes when that is exactly where it should stay.

TRACTORS AND VINES

(For Those Who Either Romanticize Nature or Glorify Progress)

But when their prosperity arises from the goodness of the climate and fertility of the soil, I partake of their happiness it is true, yet stay but a little while with them, as they exhibit nothing but what is natural and common. On the contrary, when I meet with barren spots fertilized, grass growing where none grew before, grain gathered from fields which had hitherto produced nothing better than brambles, dwellings raised where no building materials were to be found, wealth acquired by the most uncommon means—there I pause to dwell on the favorite object of my speculative inquiries.

—J. Hector St. John de Crèvecoeur,
Letters from an American Farmer

How can man prosper and not alter nature? Can we alter nature and not destroy nature? Yet how can anyone leave nature alone and still be civilized? That dilemma is at the heart of Western notions of progress itself. By the fifth century B.C., the Greeks were both confused over, and obsessed with, this proper relationship between nature (*physis*) and culture (*nomos*).

The right-wing sophists saw human existence as a state in which nature and culture were always in a death grip, not a proper equilibrium. They felt that the inferior of society curbed the innate power of the more gifted, who, like lions in the wild, might govern all the more successfully once each could be allowed to drift to his natural and predetermined station. Their idea was a natural freedom somewhat different from what our founding fathers thought were man's "self-evident" rights.

Like those lower predators of the animal kingdom who kill and lounge without cognition, who see no need to bury their dead, who are enslaved to—or, in the sophistic view, rather liberated by—instinct, the natural order rewards the rightfully strong who are not and should not be hampered by the weak. Men like the ancient Athenian rightists Antiphon and Critias would accept natural savagery as the necessary price of letting ability, not convention, establish what we can and cannot do. True liberty, in their eyes, was allowing the toothed and the fanged of society, like the carnivores of the wild, to do pretty much as they pleased. Why, they argued, would some men be born so capable and strong if not to rule in the first place?

More sober thinkers like Euripides and Thucydides acknowledged at times this primacy of innate ability. They knew that the laws of nature—the individual's pursuit of liberty and freedom especially—cannot and should not be wholly abandoned to artificial ordinances and customs of the egalitarian *polis*. But, like Sophocles and Plato, they never forgot that if it were not for the bridles of civilized life, we would revert entirely to our natural selves, that is, beastly, savage, and destructive, not much better than the Cyclopses or Centaurs, free monsters who lived in a natural state in which brutality was the real dividend of nature. Civilization and culture, as Thucydides teaches us in the third book of his history, were precious commodities, which once abandoned were nearly impossible to reacquire. Aristotle said man was a *polis*-abiding animal; otherwise, he simply was not a civilized man.

So the Greeks' *nomos/physis* paradox was raised in a variety of contexts: the beauty of the wild versus the ugly congestion of the *polis;* nature's gifted and sometimes savage supermen checked by the envious and weaker who rely on artificial custom; the human spirit, honed and perfected by constant physical assault, at war with its enervated side that arose from artifice and affluence, and man's concern for equality versus the wild's need for liberty. Without a middle ground, disaster struck equally those whose social hubris denied their natural state (for example, Pentheus of Euripides' *Bacchae*) and their natural counterparts who did not recognize the civilizing advantages that accrued when natural impulses were tamed (as Plato and Thucydides often demonstrated in their speeches).

The best technologies and practices—Sophocles' "many wondrous things" such as medicine, agriculture, construction—used man's genius to regulate, but not to stifle, the wild. That harmony, Aristotle and Plato felt, was our true and natural state, the creation of a successful institution like the city-state of some 5,000 citizens that allows us to use nature wisely to advance a culture.

No wonder vital industrial crafts that improve our lives, especially mining and metalworking, were nevertheless distrusted by the Greeks. In excess, as we in the twentieth century have learned, they were felt to have become aberrant practices, profitmongering conducted in the dank and dark, where man became lesser, not greater, for his work that might put us in an artificial world of clay and iron. By the same token, completely wild callings—herding or hunting—that did not give enough play to man's potential, were considered to be near barbarous and precultural, and but a small step away from savagery. Both the outdoorsy Artemis and her antithesis, the smithy Hephaestus, are fierce gods, but still of a lesser order than Demeter, Apollo, and Athena—the true immortals, who honor the steady progress of those who wisely and with measure master the wild.

To the Greeks and Romans the lily-white northern Europeans were wild and murderous, since their climate was harsh and gave no respite, no chance to vote and argue outdoors, visit the open-air agoras, wrestle naked, or farm olive trees and vines. Survival there demanded savagery equal to the cold. But to the east in Asia the climate was too temperate and so made soft men who could live too easily. Only around the northern shore of the Mediterranean, the Greeks believed, did its cold but brief winters and its hot summers allow man to live outdoors with nature, achieving the culture of the *polis* through the hard-fought mastery of stone, soil, and water. In this broader context, Mediterranean agriculture—vines and trees especially—took on special meaning as the best nursery of the citizen, "the best tester of good and bad men," the best mechanism to keep civilization both natural and civilized, the only place man used muscle and brain, human and natural resource, to produce in proper measure food from the ground.

We, in the richest farming area in the world, have forgotten all this, and so forgotten really what farming is and what it is for. The ecologists say it should be a natural practice (*physis*). But it is not. The agribusinessman claims for it a mechanical or scientific pedigree (*technê*). That is only half true. Out of this misunderstanding of what agriculture is comes a greater ignorance of how we should use, though not ruin, nature, how we might fabricate without denuding, how the brain without the body's muscles is no brain at all—how, in other words, we are to live.

In short, farming's lesson is simply *middleness*, which is at the heart of civilization itself. No surprise that those rare societies with three classes, not two, had vibrant agrarians—Aristotle's middling farmers who owned and lived on and passed on to their children their own land. These Greeks were not renters, serfs, sharecroppers, peasants, or tenants, overseen by a palace, castle, or enclave of clerks; they were not hired mercenaries or jackbooted militarists, but deadly and reluctant militiamen; they were not docile subjects or a capricious mob, but skeptical and tightfisted, property-owning voters.

A man of the soil, then, does not read all day, nor does he lift and hammer on specification from an overlord, his muscles in service to someone else's thought. The chauvinistic farmer thinks, exerts, reads, acts, on his own, without orders, without becoming either a brute or a sophisticated nasal-voiced doormat. Instinctively the yeoman is uneasy with the hunter, who is wild, and the mechanic, who is unnatural. He meets both daily. At times he must be both. In between the two, he comes to learn of and appreciate his niche, this rare balance between surrender to and repression of nature, between mindless physicality and sterile thought.

This quest for the mean consumes the farmer and is a war of the self. When the farmer reads in the evening, he understands that he cannot continue enchanted with a favorite novel on into the next day. Should he irrigate in freedom and without care in the late evening for a week, he is soon dour and concerned that his mind is now in more need of the abstract. He can talk well with his lawyer because he and his hired man converse as they change the tractor generator; and the hired man listens because the farmer can talk to his lawyer. Because he can wade into an orchard and stand down a rebellious crew boss, he can advance the crew boss's impoverished cousin a week's wages. The farmer pushes himself so because he knows that the easy way of doing only a single thing superbly ensures that all other tasks will be unsatisfactory; the far more difficult challenge of doing many things well, he admits, guarantees that he shall not be flawless at any. The physical effort to plant and cultivate is matched by the mental exertion to plant what and cultivate how; the farmer, who has no boss, no confidant, no counselor or advisor, then knows he must do all things himself and do them well enough, frustrated that he cannot master any. The farmer strives for the mean around him because he is at constant war with himself to avoid what feels easy, comfortable, and accustomed; people see him in town as brooding, detached, and often strange in his manner; little do they know of the battle inside his brain to plant plums, but research the right variety, to joust with the accountant but also with the grape picker, to check the price of peaches on the Internet but also weld the broken disk, to raise his children to bar forcibly the door to the intruder, but on occasion to hire the needy when they are not needed. In no other profession in America does the self strive to bridge the natural and cultural divide, to combine the brutal and refined, to work alone amid others.

The true farmer must seek technology and the progress of the ages in order to tame, but not to extinguish, the wild, from which he is nourished even as he exploits it. In our Industrial Age he uses metals and mechanical powers to shape and mold nature, but not to such a degree that he abandons the aggregate wisdom of the past when there were no such advantages, and farmers were never cut from nature's umbilical cord. And he is a captive of his past, of his grandparents and parents whose house he inhabits, whose trees he farms, whose creed he follows, and whom he also never quite escapes. He must never insulate himself entirely from their natural and pretechnological expertise—pruning, grafting, irrigation, weed removal—gained from firsthand observation of cultivated plants before the time of industry. When the farmer at last finds this proper balance between past and present—that is all that farming really is—he is loathe to give it up. He is king on his tiny plot. He now needs no one else to give him worth. He fears none who cannot take anything from him. He is Mr. Crèvecoeur's new man, still new.

I grew up with my great-grandfather's and grandfather's Thompson vineyards, worked, played, and aged amid their 45,000 vines, ninety acres planted mostly in the last century on every empty space of this farm. I have been on my knees for two years, along with my brother and cousin, planting twenty more acres of Thompsons, weeding around the young vines, staking and wiring the vineyard, training the young vinelets up the stake. I have torn out ten acres of my ancestors' grapes with a tractor and chain, hired a bulldozer to eradicate twenty and more acres and to stack their carcasses for burning stakes, wires, stumps, and all. I have planted, worked—and destroyed—Thompson vines most of my life on or beside a tractor. But I am only now beginning to understand that both—plant and machine—are near perfect in what they are and for doing what they are intended to do, nearly perfect too for explaining what nature is and how it is and is not mastered by man's culture. I see no reason why five hundred years from now a Thompson vineyard on this farm, a near-identical tractor to the one presently in the shed, and some distant offspring cut from the past cannot continue to work in tandem.

There are over 8,000 varieties of grapes in this world, grown in hundreds of different climates in a variety of soils. But within the United States nearly 90 percent of all commercial grape acreage is found in California. Of those 70 to 90 important varieties grown here, about half of the state's vineyards are planted in just one species, Thompson seedless grapes. Thompsons comprise 300,000 acres, nearly all of it in this Central Valley. America does not know it, but the Thompson seedless grape is the nation's most ubiquitous and important vine.

The Thompson grape is the seedless, sweet, oblong berry, light green to golden, found in the fresh produce section of the grocery store from mid-June to October, by far America's premier table grape. Dried, these grapes are the fat purple raisins sugarcoated in raisin bran. Chocolate-dipped, they are Raisinets. Without a candy surface, they are packed in Sun-Maid boxes and canisters. Crushed, their juice is the main though invisible ingredient in most generic white table wines. If not fermented into wine, but concentrated, that same juice is used as a silent sweetener in fruit drinks and sugar-tasting pop beverages. You can even find Thompson seedless grapes canned simply as "grapes" in fruit cocktails and salads. In other words, once picked, Thompsons are either shipped fresh, sent to the crusher for wine or juice, or laid on the ground to dry into raisins.

We encounter these Thompson seedless grapes everywhere in America because of the genius of one nurseryman, William Thompson. Over a century ago he engineered this new species from the older and smaller, less sweet, less hardy, less productive Sultana grape first grown in Asia Minor and southern Europe. Mr. Thompson surely knew that the San Joaquin Valley was hot, that from April to October it did not rain, that it was not uncommon to experience years of drought, that it endured entire Augusts where the temperature exceeds 100 degrees, that this basin was not and is not, in other words, a nice place to produce a fine table wine or a soft, vulnerable grape.

There are no hills or slopes in this Valley. No cool nights. No rocky and thin soils for making a good wine. Rather, there are scorching days, deep loams, and plentiful irrigation water, an equation perfect for any lowbrow fruit that prefers heat, light, nitrogen, and water, that likes to be allowed to grow and produce, not to be alternatively stressed and coddled. No wonder you can instantly distinguish by sight alone a Thompson grape grower from those who cultivate chardonnay or pinot noir. No ascots, bush jackets, or fanny packs among Thompson raisin farmers, no SUVs in the vineyard alleyways.

Once the snowpack of Sierra Nevada mountains was dammed and tapped in the late nineteenth century, water, the vine's lifeblood, was channeled into the Valley below. The Thompson grape was immediately planted nearly everywhere down the 200-mile Valley corridor, ubiquitously in this former desert from Bakersfield to Stockton, from the railroad that bisects the basin all the way east to the Sierra. Thousands of farm families in this century lived and died by the Thompson. They thrived due to its sheer powers of production, its supernatural abilities to produce fifteen tons and more of fresh grapes per single acre of vineyard—and just as often died by the ensuing surfeit when the winery, the raisin packer, the canner, the juicer, and the shipper said, "You farmers, stop planting that damn grape; we got all the Thompson Worthless we can handle and then some." You see, we planted it to superfluity because it was the perfect grape for this perfect Valley. Through the power of its production it was profitable even for the small farmer on his small acreage. As wheat was for the nineteenth-century Kansas sodbuster, so Thompsons were to us.

Our once monocrop farm was no exception. This ancestral house, my mother's and aunts' Stanford education, mine too, and sixty-six years of round-the-clock care for my crippled aunt in the living room, were all the bounty of a few good years. Back then the grape gave its all, and that all was otherwise in short supply, usually due to a war or natural disaster. In those rare years when the price is high and steady, the matchless production of Thompsons can bring in enough money to hoard quite a pile, enough capital for a family to get by for the next five or six years of inevitable ruin. By the same token, the present mortgage, my current refuge in the university, the dissolution of my family, the sickness and death it-

self of some of my kin, are in some part all due to the overabundance of this grape, whose cycle of ruin is more often the norm.
We are in year seventeen of yet another Thompson cycle of depression. Sixteen years it has been since the great raisin crash of
1983, when a ton of raisins plummeted from $1,300 and more to
under $500 in the space of six months. Such price collapses bring
reality to the romantic farmer. The raisin ruin that began in 1983
made it clear, for example, that our farm cannot support four families—it did that only briefly in the 1920s—but two, and perhaps
now only one. At best we hope in another five years that the price
of raisins might at least reach $1,000 a ton—75 percent of what it
was two decades earlier.

In turn the grape farmer himself has not been idle. His culture is
what allows the vine to give us the grape. In his great wisdom, the
viticulturist has created a match for this wondrous vine, a creation
made this century by man, not of water and carbon, but of steel
and rubber. His discovery is so ingenious that it has altered forever the way the Thompson vine itself is planted, grown, and har-
vested. I speak of the farmer's lowly vineyard tractor.

But by tractor, I do not mean the treaded behemoths that crawl
through the Midwest or the latifundia of the West Side of this Valley. Those monsters have thwarted the balance between nature and
machine, making farming easier at the expense of making it not
farming at all, raising production to new heights so that the farmer
must borrow even more to produce more food that fewer want.
Those enormities with cabs, air conditioners, radios, and ladders
are not really tractors. They are plantations on wheels—civilian
tanks, mobile homes and shops where one eats and listens to music. They are whole factories that move on treads and tracks. Their
wheels themselves tower above the farmer, and are guided by lasers

and computers as often as by the eye. They are designed not to work alongside and enhance man, but to replace him altogether.

Those machines have impressive power, it is true; they are confined to no vine or tree row, and so their stock and trade is the economy of scale, granted. But the muck, sounds, and elemental fury of iron hitting the soil is lost, and with its abandonment is given up the farmer's firsthand expertise of cultivation itself, his nose, ears, and mouth now blocked from the flow of the dust and the power of metal in dirt. The driver of those steel elephants becomes but a skilled mahout, no different from the brave crane operator or tough independent truck driver who does what he is told. All affinity with the man behind the plough is gone.

By tractor I mean instead open-air vineyard tractors of sixty to seventy horsepower, about eight feet wide with only four wheels, two small, two much bigger, that are designed to cultivate down a twelve-foot row of vines, the proper combination of man and machine, the smelly, visible, and precise reminder of the struggle between human skill and natural resistance. These species are, it is true, loud and powered to meet their tasks, but there is no doubt that their drivers in the vineyard are in a vineyard, not in a mobile lab, nor in a wheeled lunchroom. The farmer smells the soil he rips, and sees, feels even, the weeds he grinds up; he usually owns the soil he cultivates.

If one reads all fifteen books of the ancient agronomist Theophrastus, peruses too the Roman viticulturists Columella, Varro, and the elder Cato, even ploughs through the horrific labyrinth of trivia from that yet untranslated Byzantine agricultural treatise the *Geoponica,* he can recognize everything we now do in this Valley's vineyards—with one exception. The ancients knew all of our pests. They had poisons and fertilizers in abundance. Through their uncanny powers of observation, documentation, logic, and deduction, they could synthesize regimens of cultivation and husbandry to defy weather, climate, and animal. Even without the microscope or the canons of modern plant physiology, our empirical Greeks and Romans still fathomed the life cycle of the vine and its response to heat, water, and earth.

What, then, did they lack? The internal combustion engine. It was the one implement that would have eliminated their frightening excursus into slave breeding and control, and made their entire intricate science of basketry, animal husbandry, and human exertion redundant—if only they had harnessed the formidable powers of their intellect to the utilitarian task of mechanically pulling, scraping, and ploughing. Hero of Alexandria built a steam-driven sphere that rotated at high speeds, but no one apparently thought of harnessing such motion to a crankshaft or axle.

The vineyard tractor does not quite invert the canons of the universe—the enemies of agriculture still blow off, eat, freeze, and rot fruit. There are in this world unchanging and absolute adversaries immune to technological onslaught. The farm proves to us that enemies never cease their efforts to destroy us. But the tractor—with its instantaneous creation of hundreds of strong arms and backs, with its abrupt delivery to the ignorant farmer of the entire science of physics and engineering—widens those parameters of the struggle. It now allows man finally to craft, tame, and almost always conquer nature. The tractor is the farmer's most faithful ally in a war that he should, but cannot, give up. At present, with his tractor the farmer usually saves, rather than loses, his Thompson crop.

During its first thirty years at the turn of the century, our farm was not much different from a Greek *klêros*. Aristophanes would recognize the iron plough and harrow rusting on the hill by the shed nearby. My grandfather's wooden wagon is no different from a cart described in Hesiod's *Works and Days* of seventh-century B.C. Greece. From what I can tell from notebooks and diaries, his early cultivation regimen was not unlike Dicaeopolis' in Aristophanes' *Acharnians*. Vineyard technique and the use of animals remained largely one with the Greeks for two-and-a-half millennia—until the arrival this century of the tractor. This was the first and last machine that has allowed man to do what he wanted with his vines, to let his children put books rather than shovels in their hands, that gave the farmer enough leisure to doom him even as it sought to save him. The beautiful and efficient tractor, then, was both profit-maker for agriculture and slayer of agrarianism. No wonder the Amish rejected it even as the agribusinessmen expanded it to ten times its normal size.

Vineyard tractors have no cab other than a fabric umbrella in summer, or a tarp in winter along the engine to blow the polluted heat of the diesel smoke back into the face of the shivering farmer. Our vineyard tractors are the original tractors, which have not mutated into monstrosities. In the 1940s and 1950s, the grower of wheat and cotton likewise used these minuscule tractors. But where time and space were defined by vast expanses of open ground with no obstruction alongside or overhead save the horizon, size and power were to grow commensurably. Tractors soon became no longer tractors at all.

But vine rows do not change, expand, or contract; their parameters keep the tractor much the same as it was in the beginning. Rows check its girth, the size of its tires, the very expanse of its powerhouse—and the screwball ideas of America's engineers, who favor only greater mass and bulk. Of all the things that the farmer of vines has created, of all his inventions both chemical and mechanical, the tractor alone is consummate, perfect in form and function—in spirit like the Thompson farmer and his vineyard themselves. No wonder the local raisin cooperative, the university agricultural department, and the government farm extension service are both baffled and angry with such obsolescence. Each year they suggest that Thompson farmers "keep up," "innovate," and get on the "cutting edge" of research—only after their often ridiculous experiments, both organic and mechanical, fail to proffer any substitute for either Thompsons or pedestrian sixty-horsepower tractors. How can they?

The vineyard tractor alone, not the spray rig, not the tree topper, not the forklift, much less the poisons, and surely not the computers, can match the beauty and power of the vine to which it is paired. People on a farm still know this. Their tractors are far older than their children, often of greater age than themselves as well. Our grandfathers took this perfection of vine and tractor as a given. I think those in Europe still acknowledge the tractor's ideal unchanging form. Only in the labs and corporations of America and in its cities, where the ideal form of the tractor is unrecognized, is it constantly targeted for improvement, replacement, or obsolescence by people who don't drive it. Like the Thompson, the tractor is a product of man, but yet not a betrayal of nature.

The folk of this Valley trellis Thompsons with a pedestrian five-foot stake and single wire, or spend their inheritances in creating a vast system of arbors. They can spend $10,000 an acre and often more on such high-tensile wires, reinforced concrete posts, and high-grade steel to craft some monstrous and ostentatious espalier system for now spoiled Thompson vines. The grapes grow within such a maze too. Soon they create a continuous lush canopy for acres on end. Other growers prefer seven-foot stakes and three wires on top of each other, or still again crossarms on six-foot stakes, with two wires on the arm. Some, as I have, even try two crossarms per vine stake with five wires in all, creating double-decker trellises for the canes to climb over. Before the raisin crash, one fool bragged of such latticeworks, "Why, I have three thousand dollars an acre of iron in my vineyard"—the steel worth more than the vineyard itself when the great raisin crash of 1983 hit.

Finally you can have pretty much no trellis at all and still the Thompson grows. Our ancestors, for example, in this lumberless and steelless Valley grew their Thompsons without stakes or wire, in the manner of European wine grapes, as tiny bushes. So farmers stake, wire, trellis or do nothing and still get about all the grapes they can handle. There is no set way to trellis a Thompson vine because, unlike all other species of grapes, it will produce commercially almost any way it is trained—raisins, wine, juice, or fresh grapes for the table.

The same versatility is true of Thompson vine irrigation. Most still furrow the ground and run the water down between the rows. Some vineyards do it better than others. The relative levelness is what counts, that and the texture of the soil, the skill of the irrigator, and the efficiency of the water source. If those conditions are right, furrow irrigation is the best way to be sure that the farmer gets all the Sierra water he pays taxes for. If the ground is uneven or sandy, then it's an exercise in wasted effort, for the water never reaches the end of the row.

Whatever manner one chooses, it will work. The trick is not so much the delivery system, but the source, be it beneath or above the ground. There must be water in the subterranean aquifer to tap, or there has to be canal water flowing from the Sierra above. Because a Thompson cannot be overwatered but will produce grapes in accordance with the amount of moisture received, men of the Valley will lie to their neighbors, cheat their friends, and on occasion beat their kindred to get their hands on the Sierra water.

My grandfathers saw enough war just in these tiny environs—one man shot down on the streets of Selma, others maimed and crippled when someone took water for his Thompsons when he should not have. There are brothers who divide their father's eighty, only to incur lifelong enmity as each sibling strives to grab more water for his forty than his water right allows.

But even those abandoned and desiccated Thompson vineyards in the path of housing tracts do not die, despite having no drink for two years and more. Thompsons can be parched and scraggly and grow. Grow they do too, best in deep loams, where they are well watered and highly trellised. On flat land and on hills, they thrive both in vast acreages and on one-acre patches in front of the house. These vineyards may look not at all similar, may produce radically different-sized crops, may be the pride or embarrassment of their owners, but they nevertheless are all Thompsons, these strange vines that flourish in about any manner you please. Of course, the more wealthy and skilled viticulturists sometime denigrate the vine. "Almost anyone can grow Thompsons," they sneer.

They are right, almost anyone can—and does. The Thompson vine, for these reasons, is the most democratic of all California's irrigated crops. It is no accident that its growers comprise one of the last enclaves of homestead yeomen, whose acreage is small and whose houses are beside the vines they own.

But if Thompsons can produce wine, juice, fresh and dried fruits under almost any conditions, man's own cultural response, the tractor, can, must, match that versatility. A sixty-horsepower vineyard tractor pulls a nine-foot steel disk, a harrow, or a pair of furrows. Its prime task is to cultivate and prepare the ground between the vines: to contour the soil to allow water to run down the rows, and then to clean up the ensuing mess when weeds germinate. The tractor first and last is a puller. It is an iron ox that drags metal through a vineyard for a variety of channeling, digging, and cultivating tasks that allow the vines water without weeds—the age-old stuff of irrigated agriculture.

These toothed implements are raised and lowered in and out of the ground by the tractor's hydraulic lift. This hoist is a wondrous invention of some sixty years' vintage now. But it does far more than lift and drop the metal cultivator behind you when you prefer not to tear apart your road or garage floor. It is, of course, really a pump—so an entire array of accessory devices can be hooked up to the vineyard tractor and likewise be energized by the power of its hydraulics. Hydraulics can power a sprayer. The tractor's pressurized oil system now turns the blades of a pump under extreme pressure so that you may douse weeds, spray-paint your house, hose down your shed, or pump the feces out of your family cesspool.

Substitute a motor for a pump, and the tractor's hydraulics now run the whirling blades of the vine cutter, razor-sharp spinning knives that cut the canes before the tractor's tires and disk tear and pull them, breaking wire and stakes and shattering bunches. If dragging dead steel cultivators behind the tractor is a nineteenth-century enterprise—combustible power merely substituting for the yoke—then to meet the times you can automate your staid blades dragged behind with hydraulic whirling rototillers. There are also motorized berm sweepers, berm diggers, and brush shredders that crush, spin, and obliterate anything in their path. While they are pretty, expensive, ingenious, and testaments to the imagination of university and corporate engineering departments, they are generally not needed.

But there is far greater power still within the lowly vineyard tractor than its hydraulics. Its very power plant can do a little more than drag steel or pressurize oils, for there are ways the farmer knows to tap directly all the strength that accrues from the tractor's pistons themselves. The PTO (power takeoff) is the driveshaft, which (unlike your car's) protrudes from the front and the rear of the tractor so that the farmer may tap fore and aft into its revolutions through gears, belts, and pulleys. Blowers behind the tractor can be turned by this shaft to dust sulfur onto the vines. Far larger 500-gallon sprayers are pulled behind the tractor—even as the PTO turns finned agitators in their tanks, even as simultaneously the same PTO shaft spins a conglomeration of pulleys that turn a pump powerful enough to send out twenty nozzles full of poison, each at 200 pounds per square inch.

This utilitarian tractor can even sit still, yet aid the farmer. With the full power of the stationary tractor now turned over from its wheels to its shaft, its thrust tapped through the PTO alone, the viticulturist runs a radial saw to cut his stumps, pours his barn's concrete floor with a powered cement mixer, or pumps his irrigation water as his tractor idles for days, churning motionless along the ditch, sucking the water from canal to furrow. Remember, this PTO shaft of the vineyard tractor is still the raw power of sixty horses. It is rarely stopped. I have had sulfur machine shafts snapped by its power like matchsticks. I have witnessed a ten-ton mechanical scraper have its innards ripped to shreds by the shaft that turned too fast. I have been repelled by Ray Mix's ugly stub— a casualty when his sleeve wrapped around that shaft and his arm became but a noodle, ripped off at the socket, endlessly spinning on the PTO. After eyeing Mr. Mix's stub, I always wore my shirts short—and without sleeves.

I have had my own lesser, but still fearful experience with the lethal PTO. Ella, my Airedale, once followed too close to the tractor and spray rig. It is unwise for a dog to have curly thick hair on a farm in general, and in particular around such machinery. One wisp of her woolly coat in the shaft and in a millisecond Ella was thrown ten feet in the air, dead at foot five with a body and head at right angles spiraling off over the vines. Her wiry neck hairs stuck fast to the greasy shaft are all I remember of Ella still.

Power—enough to kill you, your dog, and your children—is the stock in trade of a tractor, not speed, agility, nor handling. But the machine's mechanical power and locomotion are not so insidious as the vine's. More instantaneous, dangerous, and ultimately transitory it is. It can, in the right hands, after all, be choked off. Its power source is fossil, not solar, fuel, and so does not expand and grow while you are asleep. Sheer traction and power to pull are what the farmer seeks from his machine. With long pistons, low gears, and treaded large rear tires, the tractor, as its name implies (Latin *tractor*, "the dragger"), is designed to overcome the natural resistance of soil to metal. As the spring tooth, plough, disk, shank, ripper, furrow, or harrow rips into the earth behind, the clods, hardpan, and sheer mud of the soil thwart their advance. In response, the treaded tires dig deeper to drag the metal through. Diesel explosions in the cylinders force the pistons down and up, up and down, turning cam and gear and hence the crankshaft, sending the power to be transmitted to the enormous wheels.

Seventy horsepower in a tractor is not seventy horsepower in a car, but rather channeled energy of a different sort. The old Volkswagen bug may go sixty miles per hour down the freeway, but its seventy horsepower cannot pull a light two-wheeled trailer over the mountain pass. In contrast, the tractor cannot exceed thirty in its highest gear; its might is harnessed and transferred for pulling and yanking, not for speed.

This sort of energy—slow, deliberate, plodding—when matched to treaded tires that do not slip, theoretically cannot be stopped on a farm. Its accelerator is not spring-loaded and engaged by the unreliable pressure of the jerking foot, but rather a stubborn lever near the wheel that, once revved, stays always where it is. There are no tank traps, no spiked holes, concrete walls, or wide rivers in vineyards hereabout. Other than an irrigation ditch or a concrete standpipe, there are few substantial obstacles to stop a tractor once loose.

I have seen such catastrophes occasionally, and the results are not the same as the drunken car that leaves the road in these parts. The latter, when the driver is inebriated and the speed approaches 100 mph, swerves into the vineyard, knocks down stakes and vines as the car rips through wire and canes. But usually three rows stop the onslaught, even when the automobile is a relic from the seventies with three hundred superfluous horsepower. Automobile tires quickly spin in the soft dirt. The metal wires hit the windshield. The car's impact is absorbed by the canopy of leaves, grapes, and wood. All motion ceases. Even the near-comatose driver rarely dies as his machine hits the rubbery rows.

The tractor, in contrast, does not stop. One occasionally reads of the dead farmer, whose heart ceases as his machine does not. He must be retrieved from a perpetually circling tractor as it ploughs through vine and post. My father fell asleep once. He went right through the neighbor's vineyard, and finally the canes, not the cessation of the tractor, slapped him awake. Our ninety-horsepower Allis-Chalmers with transmission problems engaged on its own in the shop. It almost took out the west wall until it was chased down (the tractor doesn't race, it churns). I once tried, foolishly, to knock down a dead tree with the tractor's bumper. Within seconds the tractor skidded up the bark and, bearlike, proceeded to go up the tree until the front tires were stuck straight up in the air as the rear tires ground up the trunk. Had not the sheer weight of the machine snapped the trunk, the tractor in seconds would have made a complete 360-degree turn as I dropped out of the seat with the tractor on my belly. A hired man once backed over my cousin. The tractor didn't notice the bump as it ground him into the mud, the driver clueless how to stop the assault. That it was November, not summer, and the ground was a half a foot deep in water, instead of hard and brittle as in July, put him into the house for a month rather than in the dirt for good. A different field man, who had never driven car or tractor, once was mysteriously put on our old Oliver to drive bins in from the orchard (such lunacy does happen on a farm). He went through a vine row and in panic jumped off the machine. It had to be run down like some errant bull.

I recite the litany of such horrors to remind you that things usually do not stop a tractor until its tires are flat or fuel gone. It cares little whether it crushes bone or smashes wood, whether it is sometimes a funny innocuous stray or a steel butcher, with mangled limbs and shattered skulls in its wake. Most impressive and ghastly are the rare car-tractor collisions, those occasions when the speeding auto slams into the lumbering tractor by the roadside—they are the macabre and ultimate collision between town and country. Yes, the farmer is always killed (how could he not be when perched in the open air without restraints?), but usually his machine in recompense for the crime takes the entire carload with him. The tow truck finds the car an accordion, the tractor still churning, immobile only for its blown rubber tires, its driver in pieces on the pavement. I learned, then, on this farm that the tractor needs man behind the wheel, just as the vine needs pruners— otherwise the power of both is unleashed for naught.

I have chained trees to the lift and seen the tractor rip them right out of the ground, or failing that, have had the chain rip to shreds the very stump itself. In such moments the family tractor that pulls your vineyard wagon and seems harmless in your daughter's hands is a foreign monster. With a slightly wrong move in the now chaotic world of force and resistance, it can very likely kill, throw, bury, or disembowel you.

Ultimately the vineyard tractor is like the horse—simply harnessed power greater than man's so that he might either ride on top or pull others behind. There is an assortment of vineyard wagons and trailers that the tractor, now safe and reassuringly simple, can tow on the farm—trailers in winter to carry out stumps, pipes, and stakes; trailers at harvest for boxes and bins full of picked fruit; spring trailers too, crammed with fertilizer, props, and rope. As a mere puller of wagon and trailer, the tractor is then innocuous.

The thirteen-year-old bolts through the door, proud that he too is on the machine's seat with his cousins perched on the wagon to the rear. The machine with a trailer in tow is the child's world of the plastic tractor with bright-colored accessories, his grandfather's steel now suddenly no more dangerous than a large motorized toy. "Joe Carey is a good man with a shovel," my uncle used to say, "and he can tow you boys on the wagon around this farm all day, but never let him on any tractor dragging a disk, plough, or for that matter anything else in this vineyard." Joe Carey, he knew, with a tractor and disk could be a pretty dangerous man.

The Thompson vine offers great variety, but the farmer and his tractor are always up to the challenge. I am often confused whether we have raisins, fresh grapes, wine, juice, and canned fruit from this single vine because we can grow and harvest all of them with our vineyard tractor. Or in contrast do we have tractor-driven cutters, disks, tillers, drivers, sprayers, and dusters because we must if we are to farm the various harvests from this most wondrous vine? But then, greater minds have pondered the relationship between need and response, whether man's technology itself creates things that change his universe or is but a reaction to an existing exigency.

Thompson grapes are ageless and indestructible. Until recently our earliest vineyard on the farm was known prosaically—given the farmer's imagination—as the Old Vineyard. It was 110 years old. They say (no one knows for certain) the productive life of the Thompson is eighty years. But until we bulldozed it out, the old vineyard still produced about 70 percent of the tonnage of prime vines in their twenties and thirties. It was not yet half dead. The vineyard's roots were chomped by nematodes; its old-growth redwood stakes had sunk lower into the sand as their nineteenth-century fire-burned points decayed further each year; the trellis wire, rusted and patched, was incapable of becoming taut. But the vineyard did not die—until we bulldozed and burned it. If abandoned rather than destroyed, that vineyard—despite no water, no fertilizer, no cultivation—would still now exist. Even if it were not weeded or pruned, those century-old vines still would not have perished as they did. Another century from now they would have still been alive, perhaps twisted and gnarled like a bristlecone pine, a topic of conversation near someone's backyard barbecue.

But the Thompson vine's adherence to life is not Rasputin-like. The grape is not a fanatical, ugly life force like the Johnson grass that revives maimed and disfigured when it should not, whether it is dug, sprayed, or chopped. It is not the deadly nightshade that turns only yellow for a week when doused with the plant killer Roundup or that triples and quadruples its root when cut. Instead, the Thompson grape's indestructibility is nature's perfect expression of a cultivated plant that can take heat, little water, and almost no nutrients and still give man sugary fruit.

The harsher the climate, the sweeter the grape that emerges. All of this Valley's native and introduced flora show the consequences of their hardened existence in heat and drought: thorns, briars, poisonous flowers and fruit, foul odors, and ugly hues. The

Thompson, in contrast, in the midst of Jimsonweed, cocklebur, wild willow, thistle, tumbleweed, and Johnson grass, keeps its leaves soft and light green, its canes free of thorn and sticky ooze, its stump without sharp spikes or knotty spurs. Yet it is hardier than all those more feral and savage in the plant hierarchy.

But can the culture of the tractor match the longevity of nature, when its steel and rubber are not constantly nourished by veins and channels that pass on life-giving sustenance extracted from earth and water, when tinkering men who know gears and bearings are to be matched with that divine genius expressed through the genes of plants? After all, steel cannot be yearly renewed. Chassis and axles are not regenerated when rust and corrosion have eaten beneath the ferrous dermis.

We cannot answer this question because the modern vineyard tractor of rubber tires, hydraulics, and PTO is a creation of the 1930s and 1940s and so but sixty years young in its existence. Yet it seems safe to say that vineyard tractors of those decades, more so than those built in the ensuing years, can work for eighty years, a productive life span equal to that of the Thompson. Plant as I have a vineyard, and that same day purchase a seventy-horsepower Massey Ferguson: the odds are at least even that in half a century the machine will be cultivating the vines, both mostly unchanged in appearance for the wear of five decades.

There are two large, treaded rear tires, an axle, transmission, engine, and rudimentary steering, hydraulics, PTO, and lift. Such organs, unlike the modern car, need not be light, fuel-efficient, or aerodynamic. They only have to last. The tractor is still one with America's metal contrivances of the forties and fifties, overbuilt with steel—steel wheels, steel engine, steel starter, steel shell, alternator, and cables, steel fitted with rubber hoses, tubes, and tires.

Steel inside and out. Because tractors do not need to be stylish, do not derive their value from appearance or speed, much less through the augmentation of useless chrome, wood, or gold, American engineers and craftsmen can be turned loose to do what they do best: to overbuild something to the point of being massive, strong, durable, and practical.

Outside in the shed is a thirty-five-year-old blue Ford 4000, a tiny three-cylinder, fifty-horsepower diesel vineyard tractor. My grandfather bought it in the early sixties. It has since been rebuilt and rewired. Before the odometer—properly speaking, a chronometer, since tractor wear is measured in hours, not distance driven—went out, it had gone around twice, or well over 20,000 hours. A man would have to work eight to five every day for ten years to match its cumulative record of hourly exertion. Its past service is the physical landscape that I saw on the farm this morning. It has disked, sulfured, sprayed, towed, shredded, scraped, harrowed, and ripped every vine that is now on this place. In some years it alone kept alive the vines we farm and now tear out, and made possible all the harvests that saved and destroyed us.

Like my father, grandfather, brothers, Danny Lopez, Larry Ramos, Manuel George, and countless others now dead who passed through here the last fifty years, I have sat, smoked, talked, and gone bored silly on that tractor in those decades. I was on it at age twelve, and I was on it yesterday. I have fallen asleep on it, nearly been killed on it, and on it tried to run over a foul field man who worked for a corrupt grape packer. In my twenties and thirties I sprayed on poison with it for twelve hours a day, and reeked so bad from the day's chemical blowback that I was barred from the house; the blue tractor, stained orange, likewise exiled from the shed. In polyester I have smoothed my gravel driveway with it, like some suburban tyro flush with the idea of a real ma-

218   THE LAND WAS EVERYTHING

chine under his ample buttocks. Yet in the pantheon of farm machinery, our Ford 4000 is rather young, nothing like the neighbor's Ford Jubilee, the Ferguson down the road, or the 1930s
Ford 9N that sits in the barnyard, which has outlasted three generations of its now dead pilots. Tractors, like vines, rarely die.
Those forty-year-old cannibalized or abandoned ones are junked
usually because of freak accidents, fires, collisions, or their owners' sheer inability to obtain parts. More often they reach the tractor salvage yard because the cost of some prime replacement
part—which yearly soars far beyond inflation—makes that repair
uneconomical, though not unfeasible.

My house has a pictorial history. Kin with bib overalls stand beside vines, leaning on tractors in the 1930s. Decades later, my
cousin, twin brother, and I in diapers hold up grape bunches from
the vineyard outside this office where I write. In the late sixties,
the same, though now surly long-haired grandsons, bored silly
and eager to vacate the Valley, are captured beside that now rather
forgotten Ford 4000 with the vineyard in the background—a tractor that Rees Davis, my grandfather; William Hanson, my father;
myself; and Billy Hanson, my son, have driven, in a Thompson
vineyard planted by Cyrus and Lucy Anna Davis, father and
grandmother of Rees.

If we in this country and in this family were not about to stop
this cycle, Thompson vines and the vineyard tractor could produce grapes for the rest of the next millennium, as cheaply, as
abundantly, as sanely and safely as any method any vineyard man
will yet devise. Did not Mr. Crèvecoeur say of the agriculturist's
tools, "The great objects which an American farmer ought to have
in view are simplicity of labour and dispatch"?

The relationship between vine and tractor is static and irreplaceable. There will be no virtual grape in the next millennium.

There will be no "paradigm shift" in the vineyard as promised everywhere else in America. The farmer will not be surfing the net to bring grapes to your table. There will be no Thompson website from which you can download raisins. A Cybertractor will not do. *If* the human desires grapes, there will be Thompsons and there will be tractors, and that is about the only thing I know or wish to know for certain of the century about to come.

These things will not change. For ages the state of affairs between farmer, tractor, and vine has been one continual war. Family's quotidian duties turn these friends into bitter enemies, and so the battle continues: farmer versus vine, farmer versus tractor, and tractor versus vine. I have seen neighbors milling around a tractor chained to an ancestral stump to watch the latter. As its spinning treads ground up the road, the chain links stretched and were about to explode. The tree was heaving, and everyone was rapt to see whether tractor, road, chain, driver, or stump would give first. Root versus piston always draws a crowd.

There are things in this world that even when put to foul uses cannot remain altogether base. The vineyard of the farm repossessor is beautiful, his expropriated Case tractor worthy of similar due. The long-lost kin who causes farm liquidation liquidates vines that are better than he. We sell cars in a blink; a sudden mood brings us a new Mazda. Not so with a tractor, which often shames a man who sells it for cash. The tractor of course represents technological progress; but this vineyard machine also has a stasis of its own, reminding us that its inventors eighty years ago stumbled onto some preexisting and unchanging form waiting for its proper exploitation, its almost natural contribution to viticulture heretofore untapped. Alone with one tractor and 30,000 vines, with not one improvement in either for the next millennium, a single man could feed an entire city its raisins, its fresh grapes, and much of its wine. And he could keep feeding them until we are no more.

The farmer's tractor is nothing without the vines; his vines cannot be farmed without a tractor. Without the farmer, the tractor is a worthless immovable tool; and without the farmer, the vine is a worthless rank weed. But when the farmer mounts his loud machine and guides it agilely through his lush vineyard, agriculture comes alive, as man uses just enough culture to tame, not conquer or ignore, nature. A vineyard represents a cultivated wild, a vineyard tractor a manageable machine, the farmer a man whose thoughts come alive through his muscles. A man who can master such a powerful and dangerous implement, employ it amid leaves, canes, and bunches that would not be there without it, and use his mind to direct it, his back and arms to stay on it, learns that he needs nothing more out of this life, not a different crop, not a bigger tractor, not another job. From his war with the vine and tractor, both with and against them, the farmer at last wins his war with himself, and like his perfect machine and vines realizes that he too has at last become consummate and unchanging and for the ages.

THE LANGUAGE OF TRUTH

(For Those Who Still Appreciate Direct Speech)

A pleasing competence appears throughout our habitations. The meanest of our log-houses is a dry and comfortable habitation. Lawyer or merchant are the fairest title our towns afford; that of a farmer is the only appellation of the rural inhabitants of our country. It must take some time ere he can reconcile to our dictionary, which is but short in words of dignity and names of honour.

—J. Hector St. John de Crèvecoeur,
Letters from an American Farmer

Nearly 2,500 years ago the Greek historian Thucydides demonstrated the dangers of speech that grows distant from what we know as true, as language becomes a contrived and manipulated, not merely an "embedded," part of politics to reconstruct social reality. The conniving and duplicitous say one thing, but mean quite another—and such mendacity grows in times of turmoil and ultimately brings social collapse in its wake.

Speech, of course, as part of the necessary divide between words and things, allows a space for distortion. But language, a sometimes fragile and arbitrary medium, also has a moral role: it must approximate and communicate the world that we perceive and sense, and thus not undermine but transmit a shared reality.

We can, then, learn of the moral character of a people whose vocabulary has become elastic and flexible enough to provide cover
for dishonesty. No wonder Aristotle called those who sought to
cloud logic with words the "foul fighters."

I have learned that farmers—not philologists—of all people are
now guardians of language. I have spent now fifteen years listening
to bright but degreeless men use an ethical vocabulary—"ruined,"
"smart," "mean," "stupid," "crooked," "no-good"—fifteen years
too of hearing those stupid but academic and titled subvert the
meaning of words to mean either nothing or the opposite of what
they should—"problematic," "helpful," "discourse," "context,"
and "construct."

You, reader, have no need or desire to become familiar with the ar-
cane vocabulary of the yeoman—much less the esoteric parlance of
tree and vine farming such as "frog belly" (an immature raisin),
"sheep-nosed" (a malformed plum), or "split-pit" (a swollen and
cracked peach), or even the wider diction of irrigated agriculture itself ("alfalfa valve," "berm," "twister," "head," "landing," "bowls,"
and the like). The problem with understanding farming is not learning a vocabulary unknown to suburbia or even a cryptic slang.
Rather, it is relearning words that all of us already should know, ap-
preciating the agrarian, and hence anachronistic, notion that words
really should describe an existing reality, and not be changed—that
is, mostly softened—to promote something that is false. If you
would come out here for a time, you would relearn what Mr. Crèvecoeur chauvinistically called "the language of truth":

> I speak not through the narrow channel of a partial Ameri-
> can. I speak the language of truth, and I hope that one year
> of observation will convince you of the propriety of what I
> have said.

Classicists are trained—or at least they used to be until the beginning of the decay that set in during the late 1970s—in the admittedly Germanic and philological approach. Interested in ancient birds? Wish to write of Roman divorce? Contemplating a Ph.D. thesis on the Athenian trierarchy? Start with the vocabulary, we were told. That is, round up all the Greek and Latin nomenclature to define exactly what your topic is, and what the ancients themselves thought of it through formal expression, and why and how they used the vocabulary that they did—and then all else falls into place. I too once spent five years in a graduate school to be such a narrowly trained philologist; I can testify there are limitations to such a pedantic and rigid methodology—gravity, after all, existed before there was a word for it. But at its best the study of vocabulary can capture the spirit and moral tenor of an era.

Thus I begin an explication of this misunderstood world of American family farming with a small sampling of what its words—its *glossae rusticae*—really mean, working on the old assumption that you can learn the history of a lost culture through its philology. And I also make a plea that we Americans, who have for the most part lost our art, our literature, and surely our once wonderful universities, not abandon our language, the American idiom whose frankness and honesty were once models for the world. Once one understands the clarity of agrarian diction concerning things on the farm, he can begin to understand, rather than object to, the absolutism of the farmer's vocabulary when applied to the modern world at large.

Let us start, then, with a few of the words the farmer has defined to capture what he sees, hears, thinks, does, and experiences within the tiny confines of the farm itself.

apple—rarely purchased if left unsprayed; always under suspicion of being sprayed when purchased.

application—euphemism used for spraying poisons on crops used by those who have never done it.

apricot—delicious and highly prized fruit that can be successfully produced in quality and quantity only about every fifth year.

barn—a majestic rural structure whose ancestral design renders it of no practical use in the present.

brown—the color of peaches when fungicides are not used.

cancer—a malignant occupational disease tied to agriculture, thought by farmers to occur between their fortieth and fiftieth years.

cane-cutters—lethal hydraulic blades intended to whack off vine canes, but just as likely to decapitate your dog.

cat—a well-fed, valued farm predator that eats no living pest.

coronary attack—a rather dramatic and common, though usually unsuccessful, method of alleviating farm debt.

culls—nutritious and delicious fruits that are not pretty or hard—and hence of no value.

cultivation—the choice between chemical pollution, mechanized dust-raising, or muscular exploitation.

daughter—usually ignored, often reappearing later in times of financial and emotional crisis to save farm through courageous means.

death—if done early, quickly, and insured, it can suddenly and dramatically alleviate farm debt.

ditch—a public irrigation canal, administered under communitarian auspices, where private disputes become felonious.

ditch-tender—an impressive coward.

dog—a noble guardian who unfortunately bites only those who will sue.

dove—a mellifluous farm bird that is slaughtered by visitors from Los Angeles and never eaten.

drums—pretty, sturdy, and apparently large utilitarian plastic poison containers that cannot be buried, disposed of, burned, or reused by the farmer.

ducks—farm residents that serve as more challenging moving targets for visitors from Los Angeles; sometimes eaten; *see also* **doves.**

dump—a vineyard or orchard in the eyes of town folks.

dust—extremely bothersome to brokers, bankers, inspectors, and other rare farm visitors who learn that it has some connection to the manner in which they profit.

duty—a fated tenure of about fifty years that involves the prevention of farm foreclosure.

experimental—any noun with that preceding adjective must be avoided at all costs; e.g., "experimental" peach, "experimental" fungicide, etc.

failure—the ability to grow food in abundance that no one wants.

farmhouse—a family heirloom often not owned outright but always repaired.

farm laborer—no one works harder, makes less, and is more forgotten in America.

feud—ultimately about, for, and over land.

fiber—a pretentious term used by cotton magnates to inflate their own value, as in "food and fiber." Never used by farmers themselves.

fifty—common age at which farmer realizes that he really will die in debt.

fig—a nontransportable delicacy prized by consumers over ninety without cash, and hence of little profit to the farmer.

fingers—few farmers have all ten by middle age.

Fresno—a once ridiculed agrarian backwater, whose general reputation declined with the mass arrival of a nonfarming populace.

frost—an efficient way to end the farming year before it has begun.

funeral—a purportedly somber occasion where questions of farm ownership can for the only time be calmly raised though not settled.

fungicide—to work properly must be carcinogenic.

gopher—an innocuous farm rodent that must be killed at all costs.

grandfather—long-dead ancestor whose daily presence precludes farm sale.

grapes—once boycotted by those who never wished to eat them; *see also* **wine.**

great-nephews—mysterious heirs from great distances, said to appear, for no ostensible reason on the farm in times of familial terminal illness or death.

green—the color of fruit when picked for shipment eastward.

grower—a term of self-approbation, used by those in California who often do not themselves grow anything.

hail—a natural mechanism of fruit removal that usually arrives immediately after spring plum thinning; can make the fool a genius, the genius a fool in ten minutes.

hardware store—an overpriced and distasteful local market whose viability is a barometer of the farmer's own.

harvest—a time when expectations are gradually shown to have exceeded reality.

hawk—a majestic, soaring airborne creature usually shot by age three by someone from town who is never arrested, and if arrested, never prosecuted.

herbicide—an effective chemical weed control that cuts down on dust and eventually poisons the water table.

hired man—a surly, unpleasant, and unhappy employee who is irreplaceable.

honest—a sometimes patronizing adjective used by bankers and brokers of farmers to explain their agricultural failure; the opposite of "sharp."

honor—the choice not to sell the farm; *see also* **stupidity.**

horseback rider—a more mobile lawbreaker.

in-law—an often polite stranger, who, due to marriage, divorce, or death, usually becomes co-owner of your grandfather's farm.

irrigation—procedure for increasing weeds.

jeep—a practical and utilitarian all-terrain vehicle, ideal for use in vineyards and orchards and thus in demand by suburbanites but never by farmers.

job—generally a nonfarm occupation held in little repute that earns money to be lost on the farm.

ladder—three-legged device designed to cause permanent injury at middle age; a disaster with which every tree farmer has a rendezvous.

land—is everything.

lazy—an archaic and impolite term no longer used by anybody other than farmers of those who will not work when they are able.

loan—a polite term for "loss of ownership."

mailbox—a rural roadside and federally protected receptacle, most often looted or used for target practice by the urban carbound.

material—euphemism for farm poison employed by those in agribusiness who have never used pesticides.

methyl bromide—toxic soil fumigant that destroys all living tissue and thus allows any crop to be planted where it should not be.

minimum wage—an enlightened mechanism to ensure guaranteed minimal recompense to everyone but the farmer.

move—a concept little understood by the farmer, though often promised to his children.

nectarine—a peachlike, fuzzless fruit purchased on the assumption that it is not a peach.

neighbor—a measure by which to assess one's own agricultural failure.

nephew/niece/cousin—friendly relatives who would like to be bought out with borrowed money; *see also* **liquidity**.

"next year"—a phrase symptomatic of widespread delusion psychosis, in which persistent and proven agrarian failure is fantasized into guaranteed success.

orange—the color of skin that has been accidentally sprayed.

organic—a natural method of farming that has not yet discovered how to make fruit look like it has been grown with chemicals when it has not.

organochlorides—they do not look or smell organically anything.

organophosphates—especially strange-smelling poisons, ranging from the fragrant to the reeking, known to sicken humans and save fruit at the price of contamination.

owe—in monetary terms, not applicable to relatives.

palm—natural and reusable scratch pad, instantly available for computing upcoming farm debt.

Parathion—a cheap, effective pesticide whose use on each occasion entails the death of some warm-blooded animal.

peach—a nutritious natural snack, now often shunned by consumers for its still-bothersome tendency to decay and grow soft.

persimmon—a beautiful, nutritious fruit that is always commended but rarely eaten. Seen more in paintings than on trees.

pest—anything other than a farmer that breathes in an orchard.

pesticide—poison used to ensure that fruits and vegetables are firm, attractive, and tasteless at distances 3,000 miles from their point of origin.

plum—a fruit of recognized quality, rarely purchased any longer, but in the abstract serving as a synonym for general excellence.

pomegranate—a bright-colored and tasteful fruit that requires almost no outlay, yet cannot be grown at a profit. None know how to eat it; also seen more in art than on trees.

protective equipment—bothersome clothing that ensures accidents as often as it prevents contamination.

pruning—the systematic removal of 90 percent of the past year's vine and tree growth on the assumption that sufficient fruitful buds still remain and will not be damaged by the ensuing eight months of hail, frost, and rain.

pruning shears—expensive tree and vine tools usually provided by the farmer to the worker when they are by convention and tradition assumed not to be.

PTO shaft—a tractor's rotating human-limb remover of considerable efficiency.

quince—a tree subject to sudden bacterial death, and whose beautiful, delicious fruit cannot be eaten fresh.

rabies—a nearly nonexistent deadly disease, used as pretext for the gratuitous shooting of coyotes and stray dogs.

rain—much-needed free water that usually comes at the wrong time.

raisin—nutritious, natural candy; lauded by consumers but generally not eaten by the American public unless coated by sugar, yogurt, or chocolate.

raisin plants—do not exist.

red—the color of tags of rejection purportedly signifying defective fruit, used by packers, inspectors, and distributors as citations only in times of crop surfeit.

rent—never paid to relatives.

renter—in farming, sees no reason to pay you when no one pays him.

respirator—a rubber safety device worn over the head to prevent inhalation of toxins whose use quickly results in decreased oxygen to the brain.

shovel—a reliable herbicide that as yet requires no permit.

sleeve—a faded blue towel used to clean mucus, food, and other detritus from the face.

snake—a noble and ubiquitous vineyard denizen before the age of pesticides.

son—generally recognized as the future of the farm, though more often driven off for periods of about ten years by father.

spray nozzles—orifices that discharge toxic chemicals; known to clog about every fifth minute of use and cannot be unclogged without removing protective gloves.

spray rig—a mechanical Faust that can save your crop and you along with it, at the price of your soul.

stealing—a term that does not apply to the occasional picking of someone else's fruit; *see also* **theft**.

stupidity—the choice not to sell your farm.

sulfur—a reliable, age-old, and safe natural fungicide deemed obsolete by chemical companies about every three years.

Sunday—a once-recognized holiday that used to allow the farmer to work in unperturbed solitude.

table grapes—chemically altered fruits that are not toxic in low doses.

theft—a "problematic" term when applied to the expropriation of someone else's harvests via computer or telephone.

them—everyone beyond the property line.

town—an often misunderstood thing that creeps closer.

tractor—a vital farm implement, which, if purchased new, eventually ends up as the property of the bank. Never driven by those who own more than four.

us—a rather small circle that shrinks in times of farm indebtedness; often naively used to include those relatives who do not farm but have money.

vacation—often used by farmers to describe the ten-hour work days of all others in town.

visitor—mostly potential trouble, someone who is given fruit and thanked profusely when leaving.

vomiting—a confusing symptom indicating either flu or pesticide poisoning.

water—never enough; generally felt to be the only thing worth killing over.

weed—any plant that does not bear fruit.

what if— . . . after twenty years raisins returned to $1,300 dollars a ton, plums $15 a box, and grapes $14 a lug.

wife—island of sanity, whose Cassandra-like revelations are never heeded, though often ridiculed.

wine—a tasteful and elegant product of grapes, which are always assumed to be picked by nonunion help but are never boycotted.

work—generally recognized by farmers as applicable to agricultural labor alone.

worms—reliable, natural indicators that pesticides have not been used.

Farmers often fail when they walk into the bank door, the local insurance office, the government loan bureau, or the local school, police station, or city hall. You see they talk to those who are ignorant of farming entirely—or who are ignorant of farming entirely but who make a great deal of money from those who are not—as if they were talking to farmers. An absence of euphemism, of circumlocution, of good old-fashioned American optimism and denial gets the farmer into trouble the moment he leaves his land and drives off into a world of finance, capital, and regulation that is no longer his but in fact owns him. I have seen bankers deny loans to farmers whom they otherwise would have done business with, had they not met and heard them. After all, would you loan money to, insure a crop for, or sell a policy to a man who knew more about you than you yourself—and was not shy in letting you know that? Thus for the murky world right off the borders of his land, the farmer has his own vanishing dictionary of truth.

Agricultural Extension Service—a government entity that consists of dutiful advisors who in theory would like to farm.

appraisal—always comes in high for taxes, low for purposes of loans.

assessments—forced contributions collected by right-wing officers of left-wing agencies.

assessor—most notoriously a government property-tax adjuster, ubiquitous in time of farm prosperity, mostly unknown during agricultural depression and retrenchment.

audit—a successful mechanism to obtain additional federal revenue from those farmers who cannot afford lawyers or accountants.

banker—a commercial officer who is usually very uneasy that the farmer wishes to borrow money, pay interest, and thus enrich his company.

bankruptcy—a legal procedure that, when filed by others, costs you; when filed by you, also costs you.

banks—all that the half-educated say about them is true.

Board of Supervisors—a county governing body that can give bribery a bad name; *see also* **City Council.**

borrow—don't.

bracero—an industrious helot from Mexico; as sought after by California agribusinessmen as he is feared by California suburbanites.

broker—a well-dressed salesman who profits from food when its growers cannot; a model of fine taste in costume jewelry.

buy—an unwise move, always done with borrowed money.

capital—an aristocratic banking term purportedly synonymous with money, used mostly by farm-loan examiners in contexts of its absence. Unlike labor, apparently must be protected at all times.

capital retain—yearly deductions from growers' receipts by farming cooperatives as insurance against their inevitable financial impropriety and subsequent insolvency; seldom paid back to the farmer.

cash—always worth more than fruit or land.

City Council—in comparison makes the Board of Supervisors paragons of probity.

clerks—by nature must despise farmers.

collateral—acceptable pledge of farm capital against crop loans if worth ten times the amount borrowed.

consult—as in "consultant"; in the agricultural context generally an activity in some way involving the destruction of family farming.

consumer—a well-meaning soul who wishes natural fruit of unnatural size, color, and durability, all at an unrealistic price.

cooperative—a farming organization that successfully guarantees sales for the harvests of its members at well below market value; it rarely cooperates with anyone who farms.

corporate farm—a hivelike entity whose employees by themselves do not know how to grow food but who are generous with pens, hats, and calendars.

corporation—by now well beyond good and evil.

co-sign—as Apollo said at Delphi, the wise don't; *see also* **borrow**.

costs—set by suppliers; never disputed; see **price**.

county—an inexact term used loosely of local agencies who cite, fine, and summon the farmer when they can do so to no one else.

crop—does not count as collateral, equity, security, or property.

debt—a euphemism for renting.

deductions—indecipherable runes, used by brokers as mechanisms not to turn over to farmers the proceeds of harvest sales, always appearing on official-looking computerized spreadsheets and thus rarely questioned.

degree—often now more a certification of false knowledge than of expertise or experience.

Department of Agriculture—federal bureaucracy whose employees outnumber the number of family farmers.

depression—normal period in agriculture, though usually called a "recession" or "downturn," even by those who go broke.

"Don't worry"—colloquial peremptory warning, usually given by bankers or brokers to signify the pending termination of a loan or loss of harvest recompense; sometimes followed by a forced laugh.

easement—general theft of farmland by county government without compensation to the farmer.

East Coast—a little-known and mostly unvisited region to which fruit is shipped and from which money rarely returns.

equity—a fluctuating and often arbitrary monetary abstraction that determines how long one can farm; always determined solely by those who have more capital.

estate—during a period of mourning, a money-losing farm for a brief time is deemed valuable by lawyers, the government, and distant relatives.

family farmer—a term recently appropriated by agribusinessmen, on the theory that corporate land managers also marry and procreate.

farm—an archaic term, both as a verb and noun, now generally used to describe food production in which a family is not involved.

Farm Bureau—a national political organization dedicated to any federal agricultural policy that ignores farmers who own less than 200 acres; usually administered by those who have inherited money and land but little talent.

farmer—praised fulsomely in the abstract, avoided religiously in the concrete.

farmers' markets—a rare opportunity to eliminate the middleman for the sale of minute quantities of fruit; generally recognized by the public as agricultural petting zoos.

farm policy—as in "national farm policy"; the federal continuance of institutionalized bribery, subsidy, and welfare that has nothing to do with those who grow food themselves. Usually involves awards to the victors as a form of insult to the defeated.

Federal Land Bank—a government cooperative mortgage association, whose stringent and constant audit of its small farmers ensures that the annual bankruptcy of the magnate will not destroy the membership.

field station—a publicly subsidized agricultural laboratory, whose buildings, employees, and cars multiply in times of agricultural depression.

finance—a generic word never defined, having something to do with money in the context of its absence.

food—increasingly not entirely grown from the ground.

food stamps—subsidy given to those who will eat to surfeit and thereby reduce farm surpluses; hated by farmers.

foreclosure—an increasingly common occurrence, though never voiced in polite conversation. Sometimes called a "turnover" by brokers and agents; by others felt to be euthanasia.

free market—always championed by the astute, who understand it must not be so for them to profit.

government—in America corporate agriculture's best friend, the worst enemy of agrarianism.

"I'll get back to you"—colloquial parlance used by bankers and brokers to signify that they under no circumstances will; *see also* **"Don't worry."**

inflation—a rare catch-up time for farmers; usually cursed as satanic by their farm-state political representatives.

inheritance—an opportunity to assume farm debt and to end any ideas of lucrative employment elsewhere.

initial—a verb used by the banking industry in connection with agricultural lending; usually a humane gesture allowing farmers to abbreviate their signature on all forty pages of loan applications.

inspector—a government auditor who finds near-sexual pleasure among the buildings, employees, and records of a farm.

insurance—an expensive annual fee to protect what will never be lost, broken, or burned—usually unattainable for what will be.

insurance companies—mysterious operatives who have a strange desire and ability to purchase farmland on a vast scale; *see also* **insurance.**

interest—fee for borrowing money, whose rate soars in times of inflation but remains mostly unchanged during deflation; unlike land or crops, the cost is never negotiable.

interest-free—never is.

IRS—likes farmers.

jurisdiction—important when the government wants something; irrelevant when you do.

leveraged—legalistic adjective used to describe farmers one year before bankruptcy; sometimes an adjective of admiration used by financial analysts in their twenties and thirties of the near bankrupt in times of soaring inflation.

liquidation—rumored to be a dark tunnel with bright lights at its opposite end.

liquidity—a polite term for sale, often used by cash-strapped, landowning relatives who wish to be bought out by those poorer who still farm; *see also* **liquidation.**

mortgage—rarely shrinks, generally taken out by parents after lecturing children that they should never do so.

nonperformance—a banking term to signify the financial status of a farm's loan when the land in question is overperforming in the production of food.

officer—a generic administrator, appears on farm far more often as a banking rather than law enforcement official.

permit—first stage in a process designed to provide employment for those who otherwise would not find work.

planned development—a bureaucratic term to signify the suburbanization of agricultural land according to amended statutes and altered master plans, usually involving bribery.

postagrarian—someone in the university may invent this word in the new century.

price—set by brokers, never disputed; *see also* **costs.**

principle—largely unchanged after years of paying interest.

problematic—answer given by the younger college-educated loan officer at the bank when asked about next year's loan that he has no intention of renewing.

product—a synonym for "fruit" used by those who do not farm.

Production Credit Association—a communitarian lending association that ridicules its members for borrowing money at above market rates.

professor of agriculture—generally recognized helpful expert on farming who has never farmed.

profit—a legendary concept said to be essential to farming.

properties—euphemistic term used by real estate agents to signify foreclosed orchards and vineyards that are for sale at reduced prices, as in "We have two hot properties that will move quickly."

property taxes—never go down.

raisin industry—requires a Homer to sing of its century-long woe.

rest homes—expensive and usually subsidized housing for the aged and senile where owners of property are alone asked to pay.

retirement—an often welcome state in which one is allowed to continue farmwork without pay and to pass on farm debt.

salary—an inexact term often used by the farmer to justify his own continued borrowing when expenses exceed income; fraught with psychological implications.

salesman—a clever and articulate peddler of something you don't need or want and who smiles when lying.

secretary of agriculture—a federal or state satrap who usually has never farmed and will not complete his tenure due to reception of illegal agribusiness gratuities.

sheriff—rural law enforcement officer who arrives hours after the crime only to lecture the victim why it is nonsensical that he was called in the first place.

shippers—clever folk who know how to increase the price of fruit ten times with a single phone call and without risk. Unlike brokers, they always wear boots.

shoppers—consumers disappointed that organic fruit looks organic.

stock—forced capital deductions from production credit associations and the Federal Land Bank, which increase the rate of borrowers' interest and are of no monetary value and not transferable.

stock market—is dedicated to the principle that all inflation is pernicious except vast price rises in stock.

subdivision—the conversion of irrigated farmland to houses, resulting in a net gain of available water and hence unstoppable.

subsidization—in corporate terms, the government supplies scarce water to redundant land to grow more food that is not needed but which needs price supports.

subsidy—public agricultural largess, usually federal, given exclusively to affluent agribusinessmen who under no circumstances themselves will inhabit or work a farm.

suits—a class of overseer that appears in the penultimate stage of farm disaster. They do not smile when lying.

taxes—as Gibbon said, "All taxes, at last, fall upon agriculture."

thirty days—in agricultural finance really does mean thirty days.

titles—like degrees, are nothing and everything.

title search—a $2,000 fee to assure your banker that your fifth-generation farm a century ago really was owned by your great-great-grandfather.

union—a much feared though impotent institution that has never really hurt any farmer nor helped any farmworker.

will—mysterious documents known only by their ability to appear unexpectedly and to mutate spontaneously.

zero out—a mythical state in which farmers are paid what they are owed and thus said to start off the year with the bank debt-free.

zoning laws—are not laws, but mere suggestions whose exemption and variance enrich local officials; always praised, never enforced.

If the farmer is so valuable to a consensual society, why is he so easily displaced? If he serves as a cultural brake on the excesses of democratic capitalism, why do so few conservatives worry over his passing? And if farmers themselves are so astute and fearless, why have they not crafted unions, associations, and collective strategies to prevent their own demise? Answers abound—the farmer is by nature solitary and stubborn and only with reluctance wishes to organize. The farmer lives in an ethical world first, an economically rational one second, and thus too often farms with his heart rather than with his head. The farmer cannot separate his land from his home, and so thinks making a living is the same as living. The farmer thinks locally when the world is now global, values quality when most others wish quantity, and most of all wishes to farm, when he should pack, truck, and sell. He has mastered the science of agriculture and so produces to ruinous excess, but he has failed in the arts of finance, accounting, and taxation. All these explanations help to account for the end of agrarianism. But at the heart of the farmer's crisis lies his use of language. The words he employs and the meanings he gives them do mirror reality, and do not lie. They bother and offend not merely those who would do him harm, but those who would like to do him great good as well. In this world, for the farmer to speak the truth, and to employ a vocabulary that leaves no doubt that he speaks the truth, is to make him mute.

Why is this so? Words are now properly used first to bring us together, only second to capture what our senses tell us. The utilitarian Right sees language as a mechanism to mask the brutality of unfettered capitalism—"downsizing" for firing, "retraining" for having no value, "investment" for speculation. The therapeutic Left creates new words and changes the meaning of others to make us more content and alike—"affirmative action" for quotas, "physically challenged" for crippled, "inappropriate" for perverse. But they are more alike than different, for their shared goal is to create a uniform, classless society cemented together by a

common satiation of the appetites. But because the farmer sees little need to make people happy and is indeed rarely even around
people, his linguistic allegiance is first to what he sees, hears, and
senses. For a man who thinks the market in which he sells is not
kind and the people he deals with are neither equal nor necessarily nice, there is little reason to lie. But in America what we say and
how we say it have become far more important than what we do—
and that has become a dangerous thing for men of action and not
words.

HOW IT HAPPENED AND WHY IT MATTERED: A TWO-MINUTE SYNOPSIS

(For Those Who Like Short Explanations)

Pardon my repetitions, my wild, my trifling reflections; they pro-ceed from the agitations of my mind and the fullness of my heart; the action of thus retracing them seems to lighten the burden and to exhilarate my spirits; this is, besides, the last letter you receive from me; I would fain tell you all, though I hardly know how. Oh! In the hours, in the moments of my greatest anguish, could I in-tuitively represent to you that variety of thought which crowds on my mind, you have reason to be surprised, and to doubt of their possibility.

—J. Hector St. John de Crèvecoeur,
Letters from an American Farmer

A sympathetic radio interviewer once asked of me: *So sum up, Mr. Hanson, very briefly in the time we have left why family farming is dying and why any of us should care.* He held up two fingers to in-dicate, I thought, that I had a two-minute opportunity for exegesis

before his final station break. My response went something like
the following:

Greater wealth, increased leisure, and a longer, safer, and more
exciting life in the nineteenth century were found through the rise
of cities and industrial production. In the subsequent century, the
ensuing creation of a consumer class naturally gave rise to demands for ever greater choice and bounty. Food, now a product of
mechanization like all consumer goods, was to be packaged, advertised, shipped from afar, and produced in abundance. This un-
precedented ubiquity and availability of fruits, vegetables, meats,
and grains freed Americans from both the time and expense of
growing and then buying what they ate: food, once the age-old
bane of every man, the one commodity that for centuries governed his daily struggle, was now always to be produced by someone or something else and yet priced so as to be almost free.

*Yes, Mr. Hanson, I agree, we have seen some pretty big changes,
and we are about down to the end of today's program.*

This affordability of agricultural products was the farmer's
downfall, inaugurating a cycle of increased production and decreased prices, as farm industrialization, with its hallmarks of con-
solidation, corporatization, and vertical integration, gave us more
acres and more food per acre farmed, and more brilliant and often
sinister ways of making all produce imperishable, chemically nutritious, and aesthetically pleasing. Farmland passed acre by acre
out of the hands of the family farmer and into the hands of those
few who could do what the farmer could not. Their profit came
from shipping and presenting food to the consumer, not from
growing it; their legacy was that the purpose of farmland was
solely for producing more things to be shipped and packaged for
more people.

The ancestral idea of local farm markets, of growing produce
that tasted wonderful and shipped poorly, of counting good citizens

as part of society's harvest from the farm was not merely abandoned, but deemed idiotic. In this evolution that had but one end, we received more variety and abundance of things to eat than ever before even as we had fewer farmers than any period in the history of Western civilization—the former in part always being explicable by the fact of the latter.

It is an unfortunate habit of mankind to grow complacent with its bounty and bored with the absence of struggle once its elemental needs are assured and ensuing tastes for elective gratification satiated. We ate and were clothed well, and even became more alike, but alas, grew increasingly bored with increasingly more things—the eroding fear of religion, community, and tradition an ineffective brake on our ever more dangerous appetites. Most Americans were no longer muscular, stationary, autonomous, or in a daily struggle with nature, and they sensed that their culture suffered for it. In short, democracy—once created through bloodshed by those who owned land for the salvation of a few million others who owned land, in a society that was mostly independent, physical, poor, and rural—was now to be the unconscious entitlement of hundreds of millions who were mostly dependent, leisured, secure, and suburban. Land that was once understood to grow citizens as well as food was now to grow only food for those without land.

Gone were the farmers, whose freely incurred struggle with nature had once taught citizens to be self-reliant, shamed us to say no, and so made all of us feel big, when what we owned and possessed was rather small. Left to remain was not the farmer's spirit but his mere shadow of democracy and freedom, for his ever more free, fat, and baffled progeny.

I think that about sums it up.

And thank you!

GOOD NIGHT,
MR. CRÈVECOEUR

I drove last night right from the asphalt road onto our dirt alleyway and stopped by the old communal irrigation ditch, long ago put underground. I was at the ten-foot-diameter concrete pipe that sprouted out of the soil and marked the natural convergence below of a number of subterranean concrete pipelines, where the Sierra ditch channeled off into assorted trunk lines. My grandfather had it built in the 1950s to mark his final success. With that hydraulic interchange put beneath the dirt, all six miles of open Sierra-water ditches were now piped underground on his farm. No longer on this place were there bridges, ditch-bank weeds, or washouts to contend with. No longer did a wandering ten-year-old grandson conjure up the image of a floating corpse.

That enormous air vent, with its half-dozen screw gates, drop gates, and alfalfa valves, also directed water to five or six farms down the line. I tend to avoid it nowadays for some reason—avoid it and the memory of those farmers who used to use it.

The eighty-acre vineyard of Mr. Hoskins was to our west, forty acres of good white ash soil that subsidized another forty acres of sand that should never have been farmed. Ralph Hoskins first, and Mohinder Brahr second, farmed all eighty acres, and it was no accident that both were enormous hulks, who spent their backs and arms in the age before drip irrigation shoveling furrows and cross-checks to water that sandy hill. Both spent the money they made on their good ground to make up for what they lost on their bad. Neither reached sixty-five. To the east was Vaughn Kalderian. Like all in these parts, he too farmed good and bad soil, but unlike Hoskins and Brahr, enjoyed the sand more than his loam, deeming it a measure of his worth that through parsimony, skill, and occasional ruse and stealth that he could profit from sand when others failed from heavy soil. To the south was Harry Abe, whose vineyard was as immaculate as his yard and buildings were cluttered and confused. With his tractor and scraper, he carved out a pretty good level vineyard from what had once been scrub and pond bottom. Raul Delgado farmed vegetables to the north; in his khaki pants and shirt, pith helmet, and rubber boots he sloshed through his fifteen one-acre plots of strawberries, squash, tomatoes, and carrots, barking orders at ten pickers, a not-so-happy son, and an occasional broker who delivered his weekly sales check. That agrarian mix is what the university, I suppose, now calls diversity, though in this case it was a natural, not a state-induced, phenomenon, the logical expression of farming where a man's heart, not his skin, is drawn to the soil.

In the 1950s and 1960s, when there were still real family farms around us, the pipe marked the locus for impromptu meetings, talks, and fights. Various agrarians, ditch-tenders, and the curious sometimes congregated there, yelling often, sometimes even talking, occasionally simply resting while sitting in the shade beneath the peach tree—all as the water cascaded below through the pipes and splashed up and sometimes out this huge air vent. When it is over 100 degrees and there's a ripe peach above, the sound and mist of running water in a desert can draw a crowd. The Greeks knew that; the sacred grove, the nymphs and naiads of the secluded spring, frequent classical literature for the contrast they offer to the parched hills and swelter of the Attic and Peloponnesian countrysides.

The ditch and the standpipe brought back thoughts of far uglier, rougher, and better men than those who now are our community; to them land alone was something. Vaughn Kalderian, after the second heart operation, dropped dead when the pig valve in his heart came unglued while he was picking persimmons. Mohinder Brahr—way over his head in debt, half a heart left, torn apart by the raisin crash, his kidneys ruined by medicine—used to sleep in the alleyway next to our vineyard, and died at the hospital. Ralph Hoskins lost his bluster in an inheritance squabble; too late he learned that he never quite owned the land he farmed for half a century. His millionaire Los Angeles uncle took it and gave him nothing but a bus ticket for a lifetime of vineyard improvements. Mr. Ralph Hoskins died drooling in a postalcoholic stupor in a rest home in Oakland unknown by any in his new abode—a man who farmed every day of his life and never quite understood that he did not own the ground he stood on.

Rees Davis went out the noblest. He had a silent heart attack in July 1976 after getting turned down for his driver's license. He got up the next day eager to irrigate, fell over, and died that night in the Selma Hospital. They overdosed him on Valium—"Your grandfather is just too cantankerous"—for asking to go home too frequently. "What am I supposed to do here in this place with this prostate problem?" he growled to me on his first and last day of hospital stay in this life. At eighty-six, on his way to the field in the morning—and asleep for good that night. No rest home, no dribbling, no Pampers; just trees and vines—then the good and final night's rest before the dawn.

And the farms of these ghosts of sorrows who taught me that their land was everything? At the edge of town now, all are in production for a few years yet. The vines and trees are almost indistinguishable to the naked eye from what they were forty years before—at least until the high tide of houses washes over them at the millennium. But the California farms of Messrs. Brahr, Kalderian, Hoskins, and Davis do not grow, for good or evil, citizens of their sort. Most all of such places are either rented out to landless managers, or belong to high-school teachers or store owners, wannabe farmers like myself now with primary employment elsewhere—or more often the orchards and vineyards now are some small part of a vertically integrated corporation. Perhaps it matters little since by now it is hard to tell where the town begins and the country ends anyway, since the land is but a place to produce food. But the point that these acres no longer create eccentric, independent citizens vital to consensual government is unquestioned.

And Danny Lopez, the once migrant laborer who then worked so hard for Mr. Hoskins and Mr. Davis? Well, Danny Lopez got killed in his son José's car. The younger Lopez was drunk and pulled out in front of a semi-truck. Danny was nearly decapitated in the collision. That was the fall I went off to college, so I missed the funeral. Danny's wife signed over all the Lopez furniture to the local undertaker to pay for quite a casket. My grandparents and parents objected to the local skullduggery, to no avail—the casket resembled Pharaoh's and was the nicest item that any Lopez in the long history of the line had ever owned. I saw one once in the Cairo museum almost like it. Mr. Danny Lopez, I recall, left this farm at fifty and sleeps embalmed in that mahogany box still.

But the year before he died, he organized all twelve kids and gleaned surrounding farmers' fields after their Thompson grape harvests. The Lopezes worked fifteen days picking what the regular crews weeks earlier had skipped. Then he sold those grapes green to the winery for thirty-five dollars a ton. Thirty-five dollars

a ton—less than two cents a pound for the sweetest Thompson
seedless grapes that the vineyards of Selma could produce.
My grandfather told him to quit the gleaning. Rees Davis said,
"Danny, I tried gleaning in the thirties and there's never as many
bunches left as you think, and this year even we with the first crop
are going to be about broke." In the end, he figured the Lopezes
made about twenty cents an hour each, twenty cents an hour in
1970. "This is one hell of a country," he mumbled of their work
and of their nonexistent profit, a statement whose precise meaning I'm not sure of to this day.

But the Lopezes did take the money—there is always in America, after all, some money to be had—and bought an enormous
pig, butchered it, and strung it up on the walnut tree in our joint
yard, a trophy of sorts for the now agrarian Lopezes, the autonomous, hardworking Lopezes, who at last had worked the
fields not as laborers but as farmers of sorts—as autonomous and
independent yeomen chancing the cruel odds in the crueler
world of farming. The kids brought us over pork tacos and beer,
most of the profit from ten tons of gleaned grapes. The chain that
hung that pig for some reason is still there, grown right into the
tree bark itself. I saw it today in the yard between my brother's and
my late father's house, the old Lopez grounds. After that feast,
most of the proud Lopezes returned to entitlement and petty
crime, and of course decided not to pick grapes together for themselves or for anyone else anytime again.

The Lopezes, their failure to earn a fair return for what they
picked, and their pig were part of the farmers' tragic way, the insight that courage, sacrifice, and hard work lead not to riches, not
even to survival. They need not, if the nation this year, or last season or next fall, needs no grape juice or wine and so rightly cares
little whether Mr. Lopez for the last time in his short life valiantly

tried to gather his fragmenting familial horde for a collective, honest enterprise in gleaning an unwanted and unneeded vintage.

The Lopezes were to be ruined by welfare and not saved by work. So we give money away to the idle poor, but do not give the poor the money they earned. We allow the winery to profit obscenely, because that alone allows us to buy wine cheaply and in abundance; we allow the Lopezes the dole so that they can rest and not work for what we do not need or will not pay. And when we are all done with that easy largess, then we learn that the more we make life easy for man, the worse he can become. We do all these good and noble and necessary things that fail in this country, because we are also not quite easy with the dark brutality of the brilliant accountants at the winery and their world of absolute freedom in a free market, the alternative universe of grasping autocrats and reactionary plutocrats that wash in with the high tide of true unconstraint. Even the art of Plato could not mask that nightmarish world of the lucky or gifted let loose, who, not just content with their avarice, go on to enslave and dine on the weaker as relish. So we often combine the humanity that fails with the jungle that succeeds too well, unsure of either, loyal to neither. We do know that when the talented and strong are set free, they take when they should give—but we suspect that to stop them, it would be better to be gods than men. We in our ignorance or hubris forget we can have freedom or we can have equality, but rarely for long can we have both, and so usually we end up with neither. If we are to keep our unbridled freedom and not to live as Europeans with their vast redistributive governments, taxes, and intrusive clerks, then we must have some ethical, some moral authority to say no to what we can legally do.

The farmer, because he has seen these terrible and frightening laws transpiring in nature among his enemies from the animal and plant kingdoms—his vines, spider mites and leaf hoppers, peach twig bore, brokers, weeds, trespassers, himself even—does not like such absolutes of predictable behavior, the horror that brutality and toughness bring freedom, profit, and progress, that gift and largess often create equality, stasis, and resentment. He does not like this cruel natural farming world of which he is a part, with which he must war. He would like to be tamer than he is. He does not enjoy this ironclad law that says equality, humanity even, entails the sharing of material bounty even as it ruins the mob whose poverty once ensured that their own ignorance was not dangerous. Nor does he like the human war, where he knows there is always someone—broker, banker, or neighbor—ready, willing, and able to take his land. He does not like this choice between fairness and liberty, which is no choice other than a surrender to despair. He does not like to see the Danny Lopezes of this world glean to pad the year's final profit of a corrupt winery.

The dour Brahrs and Kalderians, who do, after all, grow things, do not like the brutality that reigns over this world, and so they can only say, "Well, look at 'em work, at least those Lopezes this year are not on food stamps." They would prefer to take $20,000 from the profit-soaked winery at the point of a gun and give it to Mr. Lopez for his noble and doomed efforts at family labor and cohesion. They would like to shut down the autocrat's distillery and in its place create a cooperative of agrarian peers who would pay people the true value of what they do. They would like to forget dividend and interest and profit itself as well, if need be; perhaps ignore the laws of supply and demand, forgetting about everything other than land and work.

But they cannot. They cannot unless they can change the very soul of man himself. They nearly alone now understand this fate of man, accept it—this doom of the universe that has no solution on this earth, this law that freedom, drive, ambition, and self-interest alone create all of what we have and yet make us become what we do not like. Yes, arrogantly the farmer carries around that awful knowledge gained from agriculture, which he does not like but must acknowledge to be true and unquestioned. He does not like to see his noble neighbor thrive in hunger through adversity, and fail in complacence through entitlement.

Sometimes I think the agrarian comes not even to like himself for what this insight from agriculture, this daily testimony to plant and animal ruthlessness has done to him, this unhappy insight that character more often comes from tragedy, virtue rarely from success. Sometimes I see why at sixty he is gnarled, and tough-tongued, and apart from this world, why he gains no delight when his peach orchard finally brings a profit, and feels no sorrow when his grapes rot on the vines a week before picking, why he is not shocked when the son he set up in business on borrowed money ends up drunk and addicted and in prison, when the daughter he sent off on more borrowed money to the university comes home pregnant and poor. When the land is everything—teacher, nourisher, expressionless witness to human folly, companion to death—people can be predictable, then disappointing, and at last even irrelevant.

Aristophanes and the farmer down the road would both know that the most crass, most materialistic, most repugnant—and most pluralistic—display of human desire in Selma was the local mall or shopping center, where thousands of empowered each hour went into hock for video games, romance novels, plastic Santa Clauses, and three-pound bags of Snickers—real democracy at last, where economic surfeit finally matched the unfettered expression of America. We at last in America understood that to have real democracy you need economic superfluity, a society so materialistic, so dynamic and productive, so laissez-faire that even the oligarchs cannot steal all the bounty that vomits forth.

And is this bleak assessment of human character and the need for its humane but firm control what the crazy Thucydides was all about? Was the last speech of old antidemocratic Socrates but the same as Mohinder Brahr's? And was it true that to Aristotle and Xenophon land was everything—that the ownership of a small plot, hard farmwork, the forced separation from the urban manifestation of the species, the war to master nature and work the ground with the back and shovel alone solved the age-old dilemma between democratic freedom and republican responsibility, between dearth and greed, autonomy and slavishness, equality and liberty, crudity and effeminacy? Mostly yes. That too was what family farming taught and what the land was for, a rather brief interlude in the ceaseless and fated progression from savagery to civilization to debauchery to boredom.

And what of us? Born into an 800-foot-square farmhouse, five of us, three boys and their parents in a single bedroom, put to work in the vineyard, surviving on a single paycheck of my struggling teaching-farming dad, and what World War II profits were still around from my aged grandfather's work to trickle down, we were better than what we became. When broke, my parents both turned to the professions. When we all went to the universities, when we abandoned what made us good and embraced what made us comfortable and secure, we lost something essential, knew we lost it and yet chose to lose. Material bounty and freedom are so much stronger incentives than sacrifice and character.

When the land was everything, family farming taught us in a world of democratic sameness that there was a place for heroism, sacrifice, and self-denial, that a man who can save his raisin crop before the rain is one with the Homeric hero, who fashions a code wherein honor is no dirty word, where failure in a noble pursuit is better than success won through cowardice, where at year's end a

masterful effort to bring in the harvest is a far better thing than a
10 percent return on a bond or stock. Those nineteenth-century
German nihilists were at least right that there is much to be said
for human will, for having around a few "men with chests."

But is that Hellenic affinity across time and space not as it
should be? Yeomen are rare in history, Greek or Selman, it matters
little. The culture they create finds them passé in but a few generations. It was no surprise that Socrates talked about the bridge between the wild and the tame vine—and no surprise that Rees
Davis, without ever reading a word of Plato, employed the same
simile. It was not odd, then, that drunken Ralph Hoskins was like
the unkind Antigone of Sophocles; no stranger was he to that wisdom of inherited absolutes he professed to guard but not to enjoy.
Rather, Greek abstractionist and American agrarian were lemmas
from the same exemplar—one erudite, the other vernacular, one
old and distinguished, the other contemporary and thus purportedly commonplace, the one written and melodic, the other oral
and indelicate. But these products of a rural and autonomous society drew their knowledge ultimately from the same experience:
the rare free man who strives to create life from his own ground,
his *klêros*, who through his vines and trees in this life is given a gift
from the divine to glimpse the world as it can be if he and his kind
but work and be left in peace—if he can learn a code from his material struggle that is not of his material world.

From these men, ancient and modern, I have learned that we of the
moment and of ourselves are nothing. We are wonderful only to the
degree that we adopt some principle, a spirit to follow, which, if it
be true, guarantees heroism but a life of unhappiness since it cares
little for what most others care a great deal. So we know that we are
born, to live, and to die for something, and that something is a will
unfathomable, beyond the limits of reason and one not satisfied by

material things, one not sanctioned by those around us, but by the spirits of those who are not with us, who remind us the land alone stays behind, that we all pass on. And thank God that it is a matter of faith, that the educated and rational, gifted and rich who have invested their minds and souls in the present muck do not have a monopoly on the answers to our brief existence. Their record of contemplation and explication in the late twentieth century, their score card at the close of an awful century, is mostly one of failure: of great genius disappointed by the lack of courage needed to live out the mundane and thankless idea.

The real lesson from the twentieth-century intellectual is that the more we know, the more we read, the more we acquire, the more we travel from our spiritual, elemental existence, the more difficult it is to speak the truth, to see the just and then follow it— to live what we profess. I, who was once a farmer and am now a professor, stand indicted with all the rest. Would that it were not true! Would that education—the great, perhaps the last great, hope of the human species—were not a double-edged sword! Professors call that, I think, the "dilemma of critical consciousness," the dilemma old Cicero pondered between Natura and Doctrina, whether virtue was in one without the other, or resided in both or neither. Old Socrates was wrong: Knowledge is not virtue. Education is not the truth. The university, the latest bible, is not the answer, but the best proof of the abject failure of our times. The soul's slow acquisition of rectitude before the grave can be had far better from the wear on the body and the wisdom of heartbreak— "pathei mathos," Aeschylus said.

Farming every day reminds us of another way to deal with the cynicism of modernism. Spiritual salvation comes from the cosmos of Mohinder Brahr and Vaughn Kalderian, where the first impression, the gut instinct, the forge of tree, back, and tire are closer to the divine, to the recognition of what is right and what is

wrong, always—the only way to deal with tragedy. Orchards will
rise; disasters will follow; but the farmer, if he has an invariable
code, will deal with catastrophes that change and mutate and
reappear as he does not. If society wants him not, his trees and
vines care little; he cares little.

These farmers are there to show us how and why we have culture, what culture and nature are and what they are not. They see
nature run wild every day; they know what it is to force food from
the unforgiving earth; they understand the dangers of both the
bountiful year and the season of dearth. Thus they rarely talk if
they cannot act; they seldom philosophize when they know they
are impotent; they do not advise for others what they cannot do
themselves, they do not advise about, but live among, the working
arms and back of America, and so they find a heroism in themselves, in their denial of what they might like to do, of what others
lesser surely would do. They, unlike almost all others in America,
feel and express no easy guilt. They know, in other words, the unchanging nature of man across time and space.

From these bleak men of sorrows a precept arises: Do no harm
is about all we can do. Work hard and hope that others do the
same with the realization that most will not. Expect others to do
the easy, the wrong thing, the thing you yourself would do if it
were not for the sheer work needed for your trees and vines that
thus bridle and shape you and alter you after years in their company. Measure yourself not against the living but against the dead,
not in light of the sorry present but of the better past, not in what
man himself has made, but what he has done with and against nature. Help your neighbor; shame and forgive—but punish—the
sinner. Forget that and you lose your life, your farm, your soul—
and others too who depend on you to do the right thing and thus
too often the tough thing. Do not be generous and moral with
someone else's time, or safety—or life. It is better to be wrong and

live the life you advocate than right, but a right that you yourself cannot and will not embrace. It is more comforting, this hard-bitten dirt shoveler of the ages has taught us, to feel the disdain rather than the ovation of the mob. A hard thing to remember the more you read, the more others think you wise, the more you do not shovel or hoe.

Brace, these agrarians whispered to me that day, for what fate, not logic, dictates, and expect it to be cruel more often than fair or nice. Be assured that your turn will come, your lot of disease, accident, death, injustice, and ruin—or worse often, surfeit and lucre. Brace for it, prepare for it even, for therein lies the chance for what little audacity is left on this earth. Those farmers who had lost their crops and their lives in silence outside Selma, California, told me that there was probably a purpose for capricious unfairness, but you won't know it in this life. In the absence of explication, seek the middle ground, fearing bounty as much as want, and do not seek rational answers in the murky realms where reason does not exist.

Mr. Crèvecoeur, you had it right—we in America did for a time create a new man nourished from a unique stew of freedom and liberty. But you had it absolutely wrong too: your new man was really not new, for he was, after all, still man himself. "New" the American man seemed simply because the man of old was plopped down in a new land under new rules not yet seen before, for the last time in the history of the planet. That freedom in a vast expanse for two centuries has now hidden the old beast within—a beast who roamed nobly and widely, but not always so wisely. And did you ever warn us, Mr. Crèvecoeur, that when the land was no longer as it was before, and thus the physical culture was not any more as it was, what was to become of this new man, now grown so terribly old and familiar? What was this new untraditional man to do when the field for his boundless energy and utter unrestraint was not millions of empty acres but miles of asphalt sprawl, when he was not a restless 20 million but a bored and smug 300 million. Better that he was never so free, never so equal, that the land was never so rich, never so open in the first place, if only to save him from what we have now become. That the land is now nothing is the real diagnosis of modern man's mysterious spiritual illness.

Last night by the irrigation standpipe I made a vow to the memory of farmers past—these men of sorrows—that for the rest of my life I shall pass on something of the wisdom that they taught, the creed of those to whom I am no longer a part. And I shall.

And then more concretely I climbed up the ladder of the standpipe and as a votive for them tried to divert a little late-season Sierra water—pure, black, and cold as any Stymphalian stream in Arcadia—into the pond basin, now long dry. When that votive torrent was unloosed, the pressure to all the other trunk lines collapsed and the entire head rushed for a final brief moment into what was once the Mr. Rees Davis Pond.

"Good night, Mr. Crèvecoeur."

I heard a cry answering me in return, but when I turned around from the ditch gate they who had always been nearby me were gone, all of them now, forever, for good, gone they and their land that was once everything.

ACKNOWLEDGMENTS

I thank my colleague at CSU Fresno, Bruce Thornton, and John Heath of Santa Clara University, for reading the entire manuscript and saving me from a number of stylistic and factual errors. Michelle McKenna, of Oxford University Press, once again kindly read various drafts and offered key suggestions about the arrangement of the chapters. My editor at The Free Press, Bruce Nichols, gave much advice about the importance of clarity and focus in these essays. Daniel Freedberg, also an editor with The Free Press, went over the entire manuscript and made numerous additional suggestions, which I have adopted almost in their entirety. Fred Wiemer did a superb job of final editing; I owe both him and especially Dan a great debt of gratitude. My wife Cara as usual made writing possible. The farmers of Selma have put up with my questions and presence. Just a few of the names of local agrarians have been changed.

Victor Davis Hanson
Hanson Farm
Selma, California